Praise for **ENTWINED**

— "These vivid, original, inventive essays took me by surprise at every turn and led me to think about even ordinary creatures in new ways. Bridget Lyons is a wonderful writer, a maverick, and a free thinker, and I enjoyed her company on every page of this fascinating book."—Sy Montgomery, author of *Secrets of the Octopus*

— "We were lucky enough to be born onto a planet with so many nifty neighbors. I've held hands with an octopus, I've hugged a bunch of trees, I've talked back and forth with owls in the night—this lovely and deep book brought back those memories and sent me outdoors in search of more!"—Bill McKibben, author of *Here Comes the Sun*

— "Filled with insight, humility, and humor, *Entwined* asks all the right questions about where we humans belong while generously allowing readers to find their own answers. Highly recommended to any reader longing to rekindle their relationship with the natural world."—Leigh Marz, coauthor of *Golden: The Power of Silence in a World of Noise*

— "Bridget Lyons is an animal on the move, a curious creature with itchy feet and a passion for the wild. Her travels are filled with adventure, inspiration, and fun, but also confusion, sadness, and rigorous self-questioning: How am I to live as an integral part of a broader community of species? How am I to live on and with this awesome earth? These essays don't offer easy answers. They offer a place in which to search."—Leath Tonino, author of *The Animal One Thousand Miles Long* and *The West Will Swallow You*

— "What scientists are finally beginning to discover Lyons intuitively knows. Other species are as magnificent as humans, intelligent in their own ways, and worth paying keen attention to. Lyons' observations and insights are as embodied as they are delightful. Through her own sensory engagement and embracing vision, in *Entwined*, we experience how we, everything, are all connected."—Nicole Walker, author of *Sustainability: A Love Story*

— "From the seafloor off the coast of Honduras where ancient sea sponges filter the waters of coral reefs, to the melting sea ice on the islands of northernmost Alaska where she helps research the endangered habitat of eider ducks, the forever intrepid Bridget Lyons models for readers a different way of engaging meaningfully with the many beings of the life world. No adventure is too odd or too arduous for Lyons as she shows how we can learn—and grow in ourselves—as we make contact with the creatures of our huge, beautiful, and threatened world. *Entangled* manages to pull off that most difficult of jobs: to be informative and true to the science and yet simultaneously touch on the great mystery of the life that is all around us."—Andy Couturier, author of *The Abundance of Less*

— "Writing and thinking as clear as the freshest stream, the brightest sunrise."—Charles Hood, author of *A Salad Only the Devil Would Eat*

ENTWINED

A WARDLAW BOOK

ENTWINED

DISPATCHES FROM THE INTERSECTION OF SPECIES

BRIDGET A.
LYONS

TEXAS A&M UNIVERSITY PRESS
COLLEGE STATION, TEXAS

♾ This paper meets the requirements of ANSI/NISO Z39.48-1992
(Permanence of Paper.)
Binding materials have been chosen for durability.
Manufactured in the United States of America.

Library of Congress Cataloging-in-Publication Data

Identifiers: LCCN: 2025005218 | ISBN 9781648432873 (paper) |

ISBN 9781648432880 (ebook)

LC record available at https://lccn.loc.gov/2025005218

For my ingenious and inspiring nonhuman neighbors

CONTENTS

PREFACE

Most mornings, the first living thing I see is the tree just outside my bedroom window. It's a towering Norfolk Island pine,
at least 120 feet tall. I greet it twice: once when I awaken, then
again after I've walked down the stairs and into the driveway.
There, I pause to admire the way its scaly needles grow into
long fingers of foliage. This unusual conifer has evolved to
flourish in places like Santa Cruz, California, where I live—
places with a lot of sun; moist, salt-laden air; sandy, acidic soil;
and periods of extreme drought. I scan its upper branches for
the red-tailed hawks, owls, and ravens who take advantage of its
lofty perspective. Then I turn onto a traffic-filled street, weave
through a hotel parking lot, and walk down a steep sidewalk to
Cowell Beach. Once my feet hit the sand, I wander the wrack
line, the part of the beach where the most recent high tide has
deposited its remnants. When I stop to examine the mounds of
giant, bull, and boa kelp, hundreds of kelp flies rise out of the
rotting masses. These tiny creatures lay their eggs among the
intertwined strands, a strategy that ensures their progeny access
to an immediate post-hatch feast. I watch the insects fly toward
the waterline, where they will serve as food for the troops of
shorebirds—plovers, godwits, sanderlings, and sandpipers—
who patrol the wet sand.

This is how I notice and appreciate the creatures I like to call
my neighbors. Some people call them "nonhumans." Others

still use antiquated terms like "lower life forms." Recently, there's been movement toward compensating for that denigration by calling them "more-than-humans." Whatever term you use to describe the estimated 8.7 million other species who share our planet with us, our lives are entwined with theirs. We all occupy and are nourished by the same land, water, and air. Our actions affect each other in both visible and invisible ways.

By some quirk of evolution, we *Homo sapiens* have developed a consciousness that appears to be unique. In many respects, it has set us apart from other species. It has enabled us to solve complex problems and radically alter our environment to meet our needs. It has also empowered us to develop sophisticated narratives about our species' mission and create a hierarchy of living things with ourselves seated comfortably on top. As far as we know, other creatures do not possess this particular variety of consciousness. But that doesn't mean they lack inherent value or deserve to be displaced from their homes. Every single species on this planet exhibits an array of amazing talents, unusual habits, fascinating skills, and ingenious adaptations that enable them to thrive in their surroundings. I am reminded of this every day when I walk out my front door— even as an urban dweller.

As humans, we are still in the midst of learning to appreciate and celebrate the diversity of our own species. Only recently has seeking equality for each and every individual become a priority for many nations and cultures, and this project hasn't fully succeeded yet. While we continue this effort, we can also take the next step: extending our concept of community to include nonhuman species as well. I've been working at this for most of my adult life, and the essays in this book are explorations of my efforts.

I live in a human body armed with human senses and driven by human emotions, and I use my human brain to think and tell stories about the place of humans in the world. As a result, when I engage with other living things, I inevitably find myself comparing my life to theirs. Some people dismiss this activity as "anthropomorphism"; I celebrate it as the seed of connection. Connection breeds care and curiosity, which, in turn, engender understanding, empathy, and, hopefully, advocacy—all of which our nonhuman neighbors need during this time we have labeled the Anthropocene, the era during which the behaviors of our species have become the most powerful force on the planet.

With this in mind, I have woven personal stories into these essays, braiding them with observation and scientific research. Looking to nonhumans to make sense of my own life is one of the ways I strengthen my bonds with them while also honoring our shared complexity. In reading this book, you will get glimpses into my specific human journey. A careful reader might be able to assemble a rough and patchy memoir from these details. But presenting the narrative of one particular human life is not my intention. I am not the center of this book; I am the prism refracting the light of the creatures I present. They are species I've been extremely fortunate to interact with—directly, in the wild, in their home spaces. They are the center of this book.

In *An Immense World*, an awe-inspiring survey of the sensory capacities of nonhuman animals, author Ed Yong says, "When we pay attention to other animals, our own world expands and deepens." It is my hope that these essays will inspire you to observe nonhuman creatures with wide-open eyes—both for your own good and for theirs.

ENTWINED

TANGLED

I spend a lot of time sitting up on my surfboard, watching the surface of the Monterey Bay for the darkened wave faces that make my heart beat faster. In between sets, I pull a strand of kelp from the water and play with it. If it's a piece of bull kelp, the kind with the single stalk that's an inch or two in diameter, I grab it somewhere below the tennis-ball-sized air bladder and whip it around like a cellulose lasso. If it's a fragment of giant kelp, I lay the more delicate array on my board to admire its alternating patterns of blades and stipes—structures that would be called leaves and stems on a terrestrial plant but earn unique names on a piece of aquatic macroalgae. Sometimes, I'll wrap a braid of giant kelp around my wrist so that its air bladders— smaller ones that resemble the bulb of a scallion—dangle like charms from my mermaid's bracelet, adding some flair to my otherwise bland black neoprene outfit. These will be the most contented moments of my day, these extended breaks from my maddening habit of assessing my own worth and the value of everything around me.

The texture of wet kelp is like nothing else I know. The obvious word to describe it would be "slimy," but that doesn't quite work for me. Kelp doesn't feel like uncooked egg or earthworm skin. It is much firmer and slicker, yet still cool and wet to the touch. Unlike fish or slugs, it leaves you with no residue. Once you have pulled a stiff stalk of the genus *Laminaria* through

your fingertips, all you can say is that it feels like kelp. It is totally alien and slightly creepy, and I love it for its uniqueness.

When I spin my board around to paddle for a wave, I fling my temporary accessory back into the water. It continues to float there, always, thanks to its pneumatocysts—the air bladders—that keep much of it on the surface. There, the kelp accesses the sun's rays to participate in photosynthesis, the action that once caused scientists to put it in the plant kingdom, despite its aquatic home and lack of a root system. When three more kingdoms were added to the world of living things, kelp was reassigned to the one called eukaryotic algae, a wildly diverse gang of organisms that includes single-celled protozoans in addition to macroalgae.

It's good to know that we can rearrange our classification of things from time to time. It gives me hope that the world might be able to reevaluate me, or that I might be able to reevaluate myself.

I live three blocks from the spot where I commune with kelp. Because I live alone and work in the gig economy, I can surf just about anytime I want to. Because I live alone and work in the gig economy, I suffer from some guilt about surfing whenever I want to. I wonder if I am being productive enough, if I am doing what society expects me to, if I am squandering the evolutionary advantages of my big brain or wasting the planetary resources required to keep me breathing.

At the dinner table, my father often talked about the importance of making a contribution to society. I was never quite sure if that contribution was supposed to be artistic or scientific or something else, and I was too intimidated to ask. I also wondered how big the contribution had to be to qualify. Did creating a single painting or poem count? Or teaching a successful class? Producing a good-looking newsletter? How about making the cashier at the supermarket smile?

I've done all of those things. Most days, I don't think they're enough.

It's hard to find information about kelp that isn't focused on all the good it does for other species. When I first started clicking on marine biology websites, I was simply trying to find out the names and characteristics of the various types of seaweed I was getting caught up in. But as they became my more regular companions, I wanted to acquire some cool facts about them. Things like "giant kelp can grow as fast as 10–12 inches per day," "kelp can stretch up to 40% of its length before it breaks," and "bull kelp's air bladders contain up to 10% carbon monoxide." I learned that giant kelp—the garland that gets wound around the leash of my surfboard—prefers calm conditions where it outcompetes its cousin for sunlight. Burly bull kelp (which I discovered is also called "bullwhip kelp"—apparently, I am not the only one to have enjoyed hurling it over my head) does better in turbulent waters. In coastal California, where I live, these two vie for predominant species status. Among kelp, that is.

Apart from this smattering of details, I was mostly inundated with information about how important kelp is to other species that sit farther up on the supposed evolutionary ladder. As a literal aquatic forest, kelp conglomerations have been documented as providing shelter for over eight hundred species of marine animals, from invertebrates and fish to the central coast's beloved sea otters and sea lions. Whether dead kelp sinks to the bottom of the ocean or washes up onshore, it becomes a source of food for other living things or nutrients for the ocean itself. And then, of course, there are the human uses. Fisheries are healthier where kelp thrives. People eat seaweed directly. We extract compounds from it that appear in shampoo, pudding, soymilk, toothpaste, fertilizer, nail polish, ketchup, and a host of other commonly used items. Colonies of kelp soften the effects of coastal shorebreak. They foster human recreational activities like diving, bird-watching, and sportfishing.

These resources told me that we should value this creature for the ways in which it serves the greater ecosystem—and, ultimately, us. But what about valuing it for just being? Does that count?

If the value of a living thing is based on how many services it provides to the larger ecosystem, I'm looking pretty worthless. I'm a forty-nine-year-old woman who has consciously chosen not to reproduce, so I've already failed to turn out one of the products I could have offered the world. I would be hard-pressed to say that I have a community depending on me, since no one on my block knows my name. And my income is low enough that I don't pay taxes. I think this last one really bothers my father—a Wall Street guy who embraces the narrative of the self-made man. He's a staunch opponent of social services and a vocal critic of anyone who, in his words, accepts hand-outs. "It's important to be a producer, an earner, a contributor," he always said. When he shouted, "Make yourself useful!" from his armchair, I assumed he was referring both to folding the laundry and to choosing a life path.

According to his definition of useful—one tied to output, income, status, and achievement—I am failing miserably. When I measure myself this way, I feel like a lone herring, lost beneath a bed of kelp, where the canopy of stipes, blades, and pneumatocysts completely obscures the sunlight.

Then again, when I picture that bed of kelp, I can't help seeing how intertwined its elements are. Together, the stalks form a network of braided organisms that creates a fertile ecosystem—one in which each strand is an important part of the whole, one in which countless other creatures can find nourishment and protection. When this community undulates hypnotically with the incoming swell, I am captivated.

It didn't take much research before I found myself drowning amid the myriad ways we threaten kelp's prospects for survival. In the process of increasing our productivity and extending our fossil-fuel-driven lifestyle to more members of our own species, we've warmed the ocean nearly 3 degrees Fahrenheit in the last few decades. Kelp thrive in water temperatures between 43 and 57 degrees Fahrenheit. Even though I surf in a five-milli-meter-thick wetsuit, I know we're above that range for most of the year. Sure, I like warmer water for my own comfort; who doesn't? But when I check the surf cam in the morning to see if it's worth carrying my board down to the stairs, I cringe when I see the numbers posted below the daily report: 59, 60, some-times even 62.

On top of that, we've dumped enough "nutrient discharge" —a.k.a., fertilizer runoff from fields, dissolved synthetic hor-mones in our wastewater, chemical effluent from factories, even the dog poop from our off-leash beach—to change the pH and composition of the water into something vastly different from what the kelp evolved to live in.

And then there's the urchin problem. Sea stars, the creatures who feed on urchins and keep their population in check, have been decimated recently by both disease and warmer water temperatures. Fewer sea stars means more urchins, and a herd of urchins can munch its way through an entire kelp forest in a matter of days. In case I forget how these phenomena are linked, the shrinking colonies of macroalgae down the street remind me.

A friend from college, a doctor, was bemoaning the extension of Medicare recently. "I just don't think we should be providing full medical benefits to people who don't contribute to the GDP," she said. "I'd like to see a system that rewards worker productivity in the future." I guess that means I wouldn't get health insurance as these uncertain days approach. Which is fine, really; sea lions and sanderlings don't have it either, to say nothing of kelp.

I love my friend, but I don't want to live in her world. And I'm not sure I do—or that any of us do. As the polar ice sheets melt and the sea level rises, reclaiming both far-flung atolls (Palao, Tuvalu, Rapa Nui) and busy urban coastal zones (Miami, New York, San Francisco), the water won't pause to compare the relative productivity of those places. Yes, residents who "contribute" more and are rewarded accordingly will be able to eat sushi long after wild fish have gone extinct. They'll be able to move inland when their seaside properties are flooded. But they will not be immune to a host of other climate change realities, among them the colossal loss of biodiversity that has already begun—the outcome that makes me tear up on a daily basis when I think about a world without Coho salmon, abalone, or bull kelp. Knowing the future will contain half the number of species I can see today crushes me, even though I have no genetic skin in the long-term game.

I suppose this is why I wrap myself in a wreath of kelp every morning I'm in the water: to remind myself that it is still here, to convince myself that we are still connected and that this connection will be enough. For a moment, I let myself hope we'll embrace the belief that all living things have inherent value. Including kelp. Including me.

I like that statistic about kelp being able to stretch 40 percent before it breaks. I keep coming back to it because it reminds me that so many organisms are resilient in ways we never see. But 419 ppm (parts per million) is a lot of CO_2. That's how much we've got in our atmosphere today. I don't know if any of us are flexible enough to accommodate that.

I'm going to need to stretch my definition of accomplishment if I'm to have any chance of valuing myself. I'm clearly not going to accomplish anything that meets my father's expectations. I'm at that point in life when I'm supposed to be peaking in my career; only, I have no real career to peak in. I'm not going to raise a family. I'm not even going to grow old with someone, since I'm already more than halfway to old on my own.

Can I come to embrace my existence as valuable simply for its tiny role in a greater web? Is the advice I offered to my housemate this morning enough? Or the way I made my lanemates laugh at swim practice? How about the smile the homeless man gave me when I asked about his bicycle, or the fact that I explained the monarch life cycle to the tourists at the butterfly grove?

The miracle of the millions of chemical reactions that combine to create and re-create me every day certainly has incalculable worth. You would think, then, that the simple accomplishment of keeping myself alive would be more than enough. From where I sit right now, though, embracing that idea is a big stretch.

Yesterday, I didn't make it out as early as I like to. I was still surfing at 7:30 a.m., when the teenagers and tourists come out in full force. "Eew, this stuff is so gross!" a girl to my right screamed, as she tossed a tangle of giant kelp at her friend. "Yeah, ick. They should clean this mess up," her companion replied, flicking it off of her wetsuit like she might a spider. I grabbed a clump floating next to me and spread it on my board; the sets were at least three or four minutes apart, so I had some time. In front of me were three long strands, their stipes zigzagging perfectly, as though designed by a master mathematician. I braided them into a messy rope, then tied the ends together to form a ring. After pulling down my hood, I placed my tiara atop my head. A minute or so later, I saw one of the girls pointing at me, and our eyes met. She looked puzzled, but not disdainful. Maybe they could value the kelp in their currency of teenage adornments. Maybe I helped them do that. Maybe that's as close as we're going to get.

Inherent value, this thing I am grasping for, is an idea as slippery as a bull kelp stalk. Just when I think I'm honoring the kelp for its existence alone, I realize that I, too, am applying my value structures to it. It benefits me by giving me something to play with in the lineup, by being quirky and beautiful in its distinctly marine way, by reminding me to stop and appreciate the diversity of nature, by providing a home for the playful sea otters I love to watch at sunrise. It looks like I'm in this mix, no matter what I do. Of course, the kelp is in the mix, too, and pulling either of us out will shift everything. We are each tiny, integral parts of this ecosystem. Is participation enough for me?

I did make it out early today, and I went around the point to the more challenging break, the one where I often feel like an imposter. I'm not sure I have the skills to surf there, so I usually hang out on the edge of the main peak, where I can cede the big waves to the better surfers and grab their leftovers from my spot on the sidelines. There's a sprawling, tangled bed of kelp in my waiting area right now. It's so thick that it grabs my fins and stops my forward movement altogether—not ideal for getting up to the speed of a wave and catching it. In order to have any hope at all of actually surfing something, I had to leave my comfort zone and creep closer to the meat of the break, where the experts line up, where any hesitancy that results from thinking you're not worthy results in either total frustration or a terrifying beat down.

I watched a handful of waves go by, passing on several steep ones that I was well positioned for, and risking getting tumbled by one giant that almost broke on my head. Just as I bobbed to the crest of the huge one, I took a deep breath and looked all around me—at the lighthouse, the wharf, the horizon. From that vantage point, all the surfers looked the same, like little amphibious creatures in smooth black skins. From up high, I could see the expansive kelp colony as well, a network of greenish brown hugging the surface of the water. Its tentacles reached toward the rocks, toward the shore, and toward me. As I spun around to start paddling, the guy behind me shouted, "Yeah, girl. Go get it!" I stood up and skittered across the face of the wave. For a few moments, I felt like an integral part of it.

SOLITUDE IN DENSITY

Every morning after checking the surf, I put on my mask and guide my cruiser bike to the eucalyptus grove. There, I look for the butterflies who once congregated in these trees, forming clumps two- or three-thousand strong against the winter winds and rains. In previous Januaries, I've stood with crowds of onlookers under sleeping chandeliers of insects, watching them shudder awake in the ascending sunlight and fly off, layer by layer, to explore the world. But not now, in this pandemic winter.

Monarch butterflies have migrated here to Santa Cruz, California, every autumn for at least one hundred years, although it's never the same individuals doing the moving. Their relocations are dictated by a five-generation cycle that stretches from the coast to the Rocky Mountains, so it's the great-great-grandchildren who return to the same set of scraggly trees as their forebears. I migrated here along this same route; only, in my case, it's been the same solitary me doing the moving—more or less, anyway.

Four years ago, when I was still living out of a storage unit, the Santa Cruz monarch population numbered around thirty thousand insects. If I visited their hangouts while I was in town, I was guaranteed to see them. At temperatures above 55 degrees Fahrenheit, they would be awake, and every visible pocket of sky would be dappled with erratic flight. Below 55 degrees, they would be sleeping, dangling from eucalyptus branches in dense two-foot-long masses. I was entranced by the fortitude of these enormous clusters, especially in light of the delicacy of a single insect. Their living arrangement inspired me; it allowed for independence during sunny afternoons but required intense community commitment to survive the dark and damp coastal chill.

At that time, I was posting pictures of my winged companions on Instagram every day, not only to celebrate that I had become obsessed with my fellow West Coast wanderers but also to make sure my far-flung friends knew I was still alive—my own version of community commitment. I had recently uprooted myself from a ski town where I had lived for thirteen years. As a single forty-something metamorphosing from a mountain girl to a surfer chick, I wasn't sure I would get a chance to create a circle of support in a whole new environment. But as I fell into new routines, numbers with 831 area codes started to accumulate in my phone, and I decided I was home.

The winter I acquired a fixed address, I participated in the Xerces Society's Annual Western Monarch Thanksgiving Count. Along with a team of binocular-bearing invertebrate enthusiasts, I spent the day staring up into the tangled canopies of six established roosting sites. Despite aching necks, exhausted eyes, and wishful thinking, we saw only five thousand of our city's iconic creatures, total—one-sixth the amount I had regularly seen just two years earlier. When I read the recently released numbers for our shelter-in-place winter, breath flew from my lungs. The socially distanced volunteers counted seven hundred insects. With that dismal tally, the creature who nudged me into making this city more than a stopover site reached what scientists call "the quasi-extinction threshold" in California. One migrant settled, and the other all but vanished.

Our increasingly warmer winters and later rains aren't conducive to the extended periods of cold-induced sleep the butterflies need; that's part of the equation. But the main causes of their absence are even more directly human-induced. We're encroaching on their overwintering sites. We're spraying chemicals on their food sources. And we've outright eliminated milkweed—the plant that serves as their birthing center, nursery, and food pantry—in places like the Silicon Valley, making it nearly impossible for the generation that leaves here to survive that first migratory leap. Life doesn't get any easier for their children as they fly eastward through subdivisions and strip malls. Our species is many layers thick in this landscape.

Even though I know I won't see any monarchs this winter, I still visit their old haunts, passing by the new signs reminding me to stay six feet from other humans along the way. I would actually be thrilled to see another person in the grove; I don't get many calls from those 831 numbers these days, since we're not supposed to be congregating with people outside of our households. I stand under the trees where they once gathered, my gaze hungrily sweeping their limbs in search of the dusky brown of folded wings, in search of reassurance. I long to watch tiny creatures clinging to each other, warding off the cold and wet as one tightly knit, perfectly still, pendulous cluster. They might be there, sleeping alone, each one clutching its own shaggy patch of tree bark. But as individuals, they're far too small to see. As individuals, they're unlikely to survive the season.

Occasionally, I spot one flying, a single fluttering silhouette against a gray-blue patch of sky. I follow it for as long as I can, tracing its trajectory with my finger until I lose it in the shadows. I stay until I feel myself shiver, maybe from the saturated air of a seaside January day, or maybe from the emptiness—of the skies, of the trees, of the city street on which I pedal home, and of the hallway where I take off my raincoat and hang up my mask.

BENEATH THE SURFACE

As the evening wore on, the humpback whale's breathing became increasingly labored. From my vantage point on the beach, about fifty yards from her fluke, I could tell she was struggling. For three days, I'd watched her swim back and forth, her spouts and their accompanying forceful bellows progressively deteriorating. On the third night, the whale's signature blowhole sound was reduced to an occasional wheeze.

I lay on the sand in my sleeping bag dissecting the music of this creature's fading breath. Each humidified sigh lasted a second or two longer than the previous one, seemingly confirming my assumption: A whale was going to die a stone's throw from our camp.

I was about to finish my fifth sea kayaking trip along the Baja California coastline, working for an outdoor school that immersed young adults in an ecosystem where serrated desert ridges plunge into deep turquoise bays. We typically spent about three weeks traveling from beach to beach, hauling around the tarps, cube-shaped jugs of water, and duffle bags of food that allowed us to explore the Gulf of California's largely uninhabited landscape. Each morning, we stuffed the entire contents of our camp into the hatched compartments of our fiberglass boats, and each afternoon, we yanked everything back out again to set up another temporary home. On this particular trip, our group had been doing this for eighteen days before landing on this beach. From it, we could almost see our

trip's final destination, a spot just north of Loreto, on Baja's Highway 1.

When we had first pulled up to this campsite, we asked the students to decide how they wanted to spend their last three days in Baja. We could make a big crossing out to Isla Coronado, a dramatic hunk of rock about two miles offshore, or we could stay put, spending the remaining time exploring adjacent bays, snorkeling, and fishing without moving camp. Much to our disappointment, the students elected to stay put. They feared that a *norte* wind might whip up and strand us on the island, forcing them to miss their flights home. It appeared that our proximity to the pickup suddenly made them eager to return to the lives they had been so eager to escape from three weeks earlier.

Just after the group voted against the crossing, one of the students spotted the back and spout of a whale. "See!" he shouted from the water's edge. "We're supposed to stay here after all. It's a sign."

I rolled my eyes at Tanner, one of my co-instructors, then hurried down to check it out. In the protected bay we had recently paddled across, I saw only disturbed water—the unusual ripples that suggest someone or something is just under the surface. Then, a charcoal-colored lump crested, and I heard a long, moist exhalation. I gasped. It was a humpback. Not an uncommon sight in January, but no matter how many whale respirations I experience in this lifetime, the intimacy of hearing a giant mammal's air exchange always takes my own breath away.

We typically saw whales from a much greater distance, identifying them by the patterned clouds emitted from their blowholes: their breaths. Contrary to popular opinion, whales don't spout water. They exhale, and the air and mucus emanating from their lungs is so warm and moist that it immediately condenses into a shape that we can see. That shape, when it comes from a humpback, is balloon-like. This is one of the facts I knew about them—one that I'd already shared with the group earlier in the course. I took the opportunity to reiterate it as I stood onshore taking advantage of what outdoor educators call "a teachable moment."

I also knew what humpbacks looked like, since they are among the most commonly spotted whale species. Our whale most likely had a fully black or dusky-gray body, I told the students. As it did laps across the cove, it periodically lifted a pectoral fin out of the water, revealing the logic behind the creature's genus name, *Megaptera*, which means "big-winged." White splotches decorated its big wings, forming patterns that, like human fingerprints, are unique to each individual and allow researchers and avid whale watchers to track and identify them.

Scientists have determined that humpback whales live in all of our planet's oceans and that each marine area has its migratory routes, some up to three thousand miles long. The North Pacific population spends its summers in Alaska, where krill—the tiny crustaceans they eat—are plentiful. But that water is too cold for whale calves, so during the winter, adults mate and give birth in the warmer, gentler waters of Hawaii or Baja's Gulf of California.

"That's some of what we know about humpbacks," I said, gazing back out toward the middle of the bay where our whale had temporarily submerged. "Quite a bit, really, given that it's not as easy to study them as it is to study other animals, since we don't—and can't—keep them in captivity."

"Gracias a Dios," one of the students said, without taking her eyes off of the water.

I nodded in agreement, keeping to myself all of the things we don't know about humpback whales. One of them was about to become painfully apparent to all of us.

After our first night on that beach, I walked down to the waterline, now a bit farther away from our camp as a result of the ebbing tide. I saw a dark spot break the surface, then heard the telltale watery exhalation. The whale was still with us. This was not a good sign. It was swimming slowly back and forth quite close to the beach, as though trying to attract our attention. It hardly needed to; half of our group already sat transfixed on the wet sand where we had pulled up the kayaks the day before. I looked at Joe, my other co-instructor.

"It's freakin' dying out there, isn't it?" he said.

I grimaced. "Sure looks like it to me. I mean, it didn't sound all that vigorous yesterday, and this morning, it sounds terrible. Why else would it hang out in water this shallow?"

Joe exhaled forcefully. "So, not only are we not going to sit here on our butts for three days. We're also gonna watch a dead whale float into our camp?" He turned and walked away. "What a way to end a course."

By late afternoon on that second day, the whale had come even closer to shore, and its exhalations had both slowed down and grown farther apart. It was still crisscrossing our little bay, but it was doing so with significantly less speed and energy.

When the whale came within a hundred yards of the beach, we prohibited the students from snorkeling and swimming. Not only was it dangerous for them to share sea space with a sick marine mammal—it might behave unpredictably—we were concerned about the effect we might have on the whale's final days or hours. We considered the stress it must have already been experiencing with twenty human beings hovering nearby, witnessing its decline. But of course, it was impossible to know what this creature was feeling. We hardly knew what we were feeling.

I told myself that this was bound to happen. Animals are constantly dying in the wilderness, and by choosing to

spend months at a time camping and traveling among them,
I increased my chances of seeing one pass. I told myself that
it was good for the students to face the reality of death in this
way, to consider their own mortality—at least briefly—as they
gazed out at the horizon. But I still had a queasy feeling in
my stomach. This was not just any animal; it was a humpback
whale, one of only about a hundred thousand in the world.
They're among the largest creatures currently alive on the
planet. They're majestic, dignified, and gentle. I didn't want
this whale to die, but even more than that, I really didn't want
it to suffer.

That night, after rolling around in my mummy bag on the
beach for half an hour, I gave up and reached for my glasses.
If I wasn't going to sleep, I might as well stare up at the V of
Taurus and fixate on Aldebaran, its prominent red giant. In the
life of that star, the total amount of time we would spend on
this beach would be a microsecond. I needed that perspective.

I'm not sure if it was the ailing giant's fading life force, the
momentum-killing decision not to go to the island, or the stale,
still air—reminding us that these windless conditions would
have been perfect for crossing to the island—that affected the
students most, but they holed up under their tarps for the better part of the next three days.

On the first day, I had mustered up some enthusiasm for
a hike up the arroyo that opened into our campsite. A couple
of students and I headed out with our cameras and plant ID
books on a mission to learn the names of three new cacti. We
did it, but we also acknowledged that walking away from the
ocean felt strange, like we were turning our backs on reality.
At the same time, Tanner ran a paddling skills clinic in the
cove around the corner. He collapsed next to our camp stove
when he returned and said, "It just feels weird to be over there

because we can't be here. We all know why we can't be here, and it's hanging over us like a storm cloud."

"The students only went with you guys to keep you from feeling rejected," Joe said. "Let's face it. They don't want to hike, and they don't want to work on boat skills. They don't want to do anything. I say we throw in the towel on trying to rally them."

So the next day, we shifted our focus to finishing novels and baking cinnamon rolls with what little flour we had left. We watched the students get out of their camp chairs and wander down to the water. They would stand there and stare out at the lethargic whale, then stroll back up and lie down under their tarps. There was no fighting it; both our group and our course were disintegrating—to the soundtrack of increasingly labored whale exhalations.

That night, Joe rolled out his whale joke collection over dinner. "What's a whale's favorite sandwich?" he asked. The other two of us glared at him while munching on our bowls of margarine-covered macaroni. "Krilled cheese!" he shouted, cackling loudly enough for the students down the beach to stand up and look over toward us. "Okay, how about this one. Why should you never make a contract with a whale?"

"I've heard this," I said. "Hang on . . . it has to do with breaching . . ."

"Yeah, yeah. 'They'll eventually breach it.' You got it." Joe paused and took a swig of water from his Nalgene bottle. "Okay, this one's the best. Two whales walk into a bar. The first whale says, 'OOOOOeeeeeeeAAAAAhhhh . . .'" What followed was about two minutes of his best humpback whale song imitation, complete with clicks, screeches, gurgles, and groans. Beautiful when coming from the animal itself; intolerable when coming from a giddy human stuck on a remote beach.

The students stood up and looked again. I held my hands over my ears until he stopped sounding like a sick Wookie.

Joe made a conspiratorial face, smirked, and said, "So, the second whale says, 'I don't get it.'"

I shook my head and offered a half-smile.

Part of the reason this joke works is because we really *don't* get it. The otherworldly music of humpback whale song has confounded scientists since they first heard it about forty years ago. Not only do we have no idea what's being communicated through the complex sequences of vocalizations, but we also don't know why whales are vocalizing in the first place. The prevailing hypothesis has been that whale song is, like bird-song, a mating call. Because only the male humpbacks sing, this might seem like an obvious conclusion. However, birds' mating calls are fairly simple, very repetitive, and quite short. Whale songs are composed of sounds ranging from high-frequency squeaks to drone-like rumbles. They're extremely variable, and they can last for up to half an hour. As a result, some scientists have begun to think the singing has something to do with navigation. Perhaps it's their form of echolocation. We simply don't know. Our worldview, which is firmly rooted in our own senses and life experiences, often gets in the way of our ability to understand why other animals do what they do.

Researchers have managed to find out that the same population of male humpbacks typically sings the same song and that the song changes from year to year. Scientists have traced the lineage of different populations' melodies and found that new themes and variations appear when groups interact. They've witnessed what can only be called "cultural evolution" as they've tracked the spread of individual phrases across the oceans. As individual whales make their epic migratory journeys, they end up swimming near whales from other populations. Just like us, they hear and steal each other's material.

I love knowing that there are sophisticated musical performances happening in the sea. And I especially love knowing that, while we have some idea *how* these concerts happen, *why* they happen totally eludes us. As someone whose job it was to explain the natural world to my students, I appreciated the moments when I had to shrug my shoulders and say, "No one knows."

The third day at that camp—our final day in the field—was dedicated to "doing evals," as we called it. The students wrote evaluations of us, we wrote evaluations of them, we wrote evaluations of the course, they wrote evaluations of the course—we stopped short of writing evaluations of their evaluations, but instructor teams frequently joked that we might as well. Not many of us enjoyed this process to begin with, but it was especially hard to do when we had spent three days sitting in one place. Did Jenny show leadership? Had Tim shown initiative? Who knew, and who cared.

Still, we did what we needed to do. Tanner, Joe, and I tried to keep the focus of our comments on the eighteen days leading up to our stagnation point. Not all of the students followed our lead. I distinctly remember one telling me that I was unsupportive of the group's decision. "She had a bad attitude about our final camp," he said, "and she seemed distracted by the whale." How could I not be when I was wondering what would happen when we woke up to a beached giant? Were we supposed to call it in? If so, to whom? Would some group of people from the university or animal control or the Loreto fire department show up with knives and ropes? Or would we just watch the vultures do what they do best? And just as significantly, would I break down and lose it when I saw a huge, lifeless eye staring back at me from a foot away?

We did all the standard end-of-course things that night: We talked about what it's like to reenter the frantic world of television, telephones, and traffic. We prompted the students to journal about what they had learned in their three weeks of wilderness travel. We even played "birdie on a perch"—a silly game they adored that involved pairs of students (one a birdie and one a perch) jumping on each other's backs when the designated caller screamed either, "Perch on a birdie!" or "Birdie on a perch!" We just did it with less energy than was the norm. The presence of the whale was inescapable.

After the three of us consumed the last of our lemonade crystals and finished commiserating about our lackluster closing meeting, I went down to the beach alone and laid my sleeping gear out on the sand. I crawled into my mummy bag and stared straight up at the empty, inky sky. I sang my way through "Hotel California" and "Dust in the Wind" to keep myself from obsessing over when—or if—the next thin breath would come.

At some point, I must have fallen asleep, since I opened my eyes to a pink-splattered sunrise, the kind of brilliant sky-wide impressionist painting we often witnessed while launching our boats for a move. It only took a second for me to snap back into the hyperalert mindset of the previous three days—the one that was closely attuned to the whale's breath.

I didn't hear it as I rubbed the salt and sand from my eyes. I didn't hear it as I lay there thinking about the breakfast of fresh milk and homemade banana bread that awaited us. I didn't hear it when I stood up to pack my gear. And I didn't hear it as I walked down to the water's edge to pee, fully expecting to stumble into a gigantic carcass along the way.

Down at the wet sand, I squatted in ankle-deep water and sighed, releasing a combination of relief and sadness into the chilly predawn air. She had gone elsewhere to die. It had been a rough way to end a trip, but it was ending. I felt no wind on my face, which meant we would have an easy paddle around the corner, putting us at the trucks within an hour. After I pulled up my shorts, I stayed where I was for another couple of minutes, scanning the pastel-smudged horizon. Behind me, I could hear rustling nylon and clanging pots, the sounds of my human companions who would soon join me by the kayaks.

I took a few deep breaths and clasped my hands behind my back to stretch my sore paddling muscles. Then, I looked out toward Isla Coronado and saw it—the familiar balloon-like spout, backlit by low-angle rays. I was surprised that the whale had been able to muster up the energy to get all the way out there. She was going for one last big swim, I thought, and I strained my eyes in an effort to spot a fin or a part of her back.

I didn't see either of those things. Instead, I saw a second spout—much smaller and thinner, but still balloon-shaped. The spout of a baby humpback whale.

My hand flew to my open mouth as I watched the pair leave the cove, the condensation of their breaths vanishing along the horizon and their bodies beneath the surface, invisible to me.

Gracias a Dios.

WHEN THE CHESTNUT FALLS FAR

My father is obsessed with the American chestnut tree. This might not seem odd if he were a biologist, an arborist, or a forester. Or if he, over the course of his eighty-two-year life, had ever taken any kind of interest in landscaping. But he's a Wall Street guy. Besides money, his passions are fly fishing, opera, and Scotch. How the possible comeback of a species that disappeared from the American landscape before he was born fits into this assemblage of interests is a bit of an enigma. It does give me—a tree-hugging former wilderness guide—something to talk about with him, though.

I ask him questions about chestnuts in a desperate attempt to understand the constellation of things he cares about. For many years, I assumed that I wasn't one of those things. I assumed that, when all was said and done, our story would be one of gradual estrangement. But I wonder, now, if a tree struggling to come back from near extinction might offer us a new ending.

The American chestnut story, until recently, has been a tragic one—not unlike the plot lines in my father's favorite operas. Prior to about 1904, the species blanketed the East Coast of the United States. Its hard, fast-growing, rot-resistant wood was just what a booming country needed to fuel its

expansion. Countless houses, floors, pieces of furniture, and railroad ties were made from the strong, straight-grained timbers it produced. Its nuts were a significant source of nutrition for humans and wildlife as well as a preslaughter fattening agent for cows and pigs. Black-and-white photos of chestnut trees show entire families posed for portraits in front of their trunks with plenty of space left on either side of Mom and Dad to give the viewer a good look at the bark. Their giant boughs offered broad-leafed shade on streets from Maine to Mississippi.

Then the blight came. All of the nearly four billion trees in the American chestnut's historic range succumbed to a fungus that arrived in the New World from Asia. *Cryphonectria parasitica*'s spores spread rapidly on the wind, invading trees through wounds in their bark. Once inside the cambium, the fungus released acids that lowered the pH of the trees to lethal levels. While the blight has not killed off chestnut trees entirely, the continued presence of this tiny organism has made it impossible for them to grow any larger than shrubs in the geographic corridor where they once thrived. These shrubs have thriving root systems and can send up enough shoots and leaves to eke out an existence, but the majestic trees of legend exist only in history.

For now, that is. The chestnut's ending hasn't been entirely written yet.

The little I know of my father's story isn't tragic at all. In fact, his tale is a classic variation on the American rags-to-riches formula. He was born the third of six children to Irish Catholic parents in Queens, New York. His father was a traveling candy salesman who lived on the road, and his mother was a brilliant woman who, after graduating from college at age eighteen, reluctantly found herself in charge of a full house. My father, the only boy of her brood, made himself scarce. After the family moved to Reading, Pennsylvania, no one noticed that he spent most of his time hunting squirrels on the fringes of the city.

Somewhere along the line, he must have done some studying because he won himself a scholarship to a Catholic college in Philadelphia. After serving his Korean War time, he returned to New York City to get an MBA, courtesy of the US Army's GI Bill. Then he made the money his parents never had. He's run banks and hedge funds, sat on corporate boards, and appeared on television talking about savings and loan crises. His parents lived long enough to witness their wayward working-class kid from Queens transform into an upstanding scion.

They didn't live to see him have grandchildren, however. My brother and I are both middle-aged and child-free, so that part of the story is in the books. I suppose the dying branch of a lineage might appear tragic to some people, including my father. It doesn't to me. I think there are other ways for family trees to grow, other ways to write the narrative of a good and meaningful life. It is on this point that I have always assumed my father and I differ.

My father and I planted two trees together when I was young. One was a Japanese maple, the other a magnolia. I distinctly remember helping my father with the excavation. "You've seen the rock walls we have lining our yard," he warned me. "They all came out of the ground when these foundations were dug. This is not going to be easy." A few fading photographs in an old family album serve to testify to the fact that I did do some digging. They also remind me that these trees were, once upon a time, only a foot or two taller than the four-foot me.

While I hit five feet eight inches in eighth grade and stayed there, these trees just kept growing throughout my childhood and beyond, with only sun, water, and soil to feed them. I think we humans, as relatively shortsighted creatures, underestimate how huge our tree companions can become. Clearly, my parents did, since they recently had to take the magnolia down when its branches started to scrape the sides of the house.

When I see the Japanese maple now—when I visit my parents to make sure they're still healthy enough to live on their own—I barely recognize it. I would have never thought I would live long enough to see it tower over the house. Maybe my father did, though. Or maybe he planted it thinking about his grandchildren.

I first learned about recent efforts to bring the American chestnut back from relative extinction from my father. He explained with astonishment that there's a nonprofit organization, the American Chestnut Foundation, dedicated entirely to this endeavor. I listened as he talked about blight-resistant hybrids and gene splicing. "And get this," he added. "I've got a plan for this group to give me some that we can plant at the club."

"The club" is a fishing and hunting property in the Pocono Mountains of eastern Pennsylvania that he and fourteen other "owner/members"—all men, all wealthy—enjoy and steward. I have a hard time stomaching multiple aspects of the club, most of them having to do with its exclusivity, but I recognize the good it's doing as well. Not only does the association preserve watershed land, deciduous forest, and stream quality; it also gives my semiretired father something to nurture. When I've visited the place, I've been reminded of the days when he went camping in the Catskills with my Girl Scout troop, before he and I both got too busy to spend time in the woods together. So when club news dominates our calls, I don't mind listening.

Lately, the updates have had less to do with trout and deer and more to do with chestnuts.

As he explained, and as I later read and reread online, genetic engineering at three different levels has created the possibility of restoring the American chestnut to eastern forests. At the most basic level of manipulation, the heritage trees have been crossed with blight-resistant Chinese chestnuts as well as other root-rot-resistant strains. The resulting tougher individuals have been planted in large swaths across the American chestnut's original home range. In addition, geneticists from the State University of New York, using CRISPR (gene editing) technology, have implanted a blight-resistant wheat gene into the chestnut's DNA, creating modified chestnuts with disease protection. Finally, work is being done to weaken the blight fungus itself. By introducing a virus into its genetic material, scientists hope to make the American chestnut's mortal enemy less formidable.

In addition to raising funds and organizing volunteers, the American Chestnut Foundation serves as a clearinghouse for much of this research. They also periodically give chestnut saplings to people like my father. "Well, they're not exactly giving them to us," he explained. "I have to pay about ten bucks per tree. And I have to prove to the organization that we can keep the deer from munching on them, which is going to require some costly fencing."

"What does the rest of the club think about this?" I asked. "Are they into it?"

"Nah. They could care less, as long as the streams stay healthy and full of fish. But they're not stopping me either."

I was impressed. This struck me as a worthwhile project, if a bit like expensive and time-consuming windmill tilting. It actually sounded like something I would do.

Sometimes, when I'm on the phone with my father, I wonder how other men his age spend their time. I suppose some play golf and eat meals with friends, others do woodworking projects in their garages or volunteer with local nonprofits, and still others learn new languages or travel. Then I remember that a lot of them spend huge chunks of time playing with their grandchildren, or making things for their grandchildren, or planning to visit their grandchildren, or FaceTiming with their grandchildren.

I've never wanted to have children. I announced this to my parents when I was young, and I'm sure they looked at me with knowing smiles and said that I would change my mind. As the significant birthdays passed—twenty, thirty, forty, and now fifty—it must have become apparent that I was sticking to my guns. To their credit, my parents have never made me feel guilty about this. Still, I once worked up the courage to ask my mother if she thought my choice bothered my father. "Oh, yes, definitely. He loves children, you know," she said. "He really likes teaching the other men's grandkids to fish at the club." I shook my head, wondering what it would have been like to grow up in a family where conversations like this happened when they could have altered the storyline. Then my mother added, "He spent a lot of time with you when you were little. You remember, I'm sure. He likes people when they're young and impressionable."

I often think about the interactions I had with my father when I was just about the same height as the trees we planted. We took our German shorthaired pointer down to the lake to run off leash, we went to the hardware store, and we raked leaves and took them the town dump. We also spent a lot of time working on my softball skills. He had been a mediocre baseball player as a teen, and I think he hoped that I would do better. I must have been about eight or nine when my father brought home the two mitts that we oiled weekly to keep them supple and foster the development of the ideal softball-shaped pocket. From the start, I had a good throwing arm, so the two of us would stand at opposite ends of the front yard—one of us alongside the Japanese maple tree, the other backed up against a giant oak—playing catch. "More follow-through," he would say. Or, "Hear that? That sound? That's a solid throw, when you hear that smack in the mitt." I knew that, and I would swell with pride when he noticed. "Now, do that again," he would say. "Imagine you've got a runner to beat." And I would. After a few years, I started pitching, and these sessions evolved into lessons on concentration and focus under pressure. All too often, my pitches would fling wildly into the air, knocking the buds off of the magnolia. I was, in short, inconsistent, and games were more stressful than fun. I quit before I had any chance to rewrite the family's history with the great American pastime.

About that same time, I felt like my father disappeared. He started to travel more frequently for work, and when he was in town, he didn't get home until after my brother and I had eaten dinner. That was part of it, but not all. I know a lot of fathers with busy schedules who manage to connect with their children. I was still doing what I was supposed to do—getting straight A's, playing clarinet, singing in the school chorus, taking art classes—but I was also spending a lot of time watching MTV and debating whether Duran Duran or Culture Club was the better band. I wore mesh shirts like Madonna, and though I wasn't allowed to date, I took a great deal of interest in my friends' romantic pursuits. I debriefed the days' events every evening on the phone with the three or four girls who formed my eighth-grade inner circle.

Maybe he didn't like who I was becoming. Or maybe he just didn't like *that* I was becoming. I was a person forming likes and dislikes of my own. I didn't need careful cultivation anymore, and I didn't seem to be growing as straight and true as he wanted.

When I was fourteen, my parents told me I wouldn't be attending the public high school up the street from our house. It was then that I discovered how much of my plot had already been devised for me. I would go to a private high school that would assure my acceptance into a prestigious college that would launch me into a career that climaxed with money and success. I'd thought I would spend my freshman year taping photos of the Go-Go's onto my locker and walking home after school with my girlfriends. Instead, I rode a bus in New York City commuter traffic to an institution where my classmates both took calculus and snorted coke by the age of sixteen. Meanwhile, I still had braces and glasses and watched *M*A*S*H* reruns on TV with my younger brother. A month into the school year, when I came home bawling and begging to rejoin my friends at the public high school, my father scowled at me. "What, and just be average? No."

Soon after, I stopped calling him "Dad" and started calling him "Father."

The American Chestnut Foundation, which was created in 1983, wasn't always on board with the genetic manipulation of trees. At first, they wanted to stick with traditional cross-breeding, the kind that's been a part of agriculture since the days of Gregor Mendel. By 1990, they had warmed up to the experimental genetic work being done by a team of scientists at the SUNY College of Environmental Science and Forestry. In 2013, this team unveiled an American chestnut variant, called Darling 58, that seemed to be resistant to the *Cryphonectria* fungus. Darling 58 still needs to be evaluated by the EPA and the FDA (because the nuts it produces will become food), so government approval of this chestnut strain will be a long and complex process. In the meantime, it's launched the chestnut version of the increasingly common GMO (genetically modified organism) debate. On the one side are the pro-modification people who argue that since our species' movements brought the blight to American shores in the first place, our species should fix the problem. They argue that engineering pervades every other aspect of our world and that we might as well use it to restore a beautiful and useful tree to its rightful habitat. On the other side of the debate, opponents of Darling 58 claim that these genetically modified trees will cross with the few natives that remain and eliminate the heirloom strand from the gene pool. They also point to the fact that Monsanto has been waltzing around the GMO chestnut scene of late and allude to the dangerous precedents that might be set by the species' adoption. If we promulgate this one superspecies, they ask, what's to stop us from redesigning entire forests with "perfect" trees, all according to the latest fashion?

"It's ridiculous that anyone would contest the introduction of this tree," my father said. "Nothing but good will come from this. Imagine: Chestnuts all up and down the East Coast again. It would be an amazing sight. What could possibly go wrong?"

I could think of a few things, but I kept my mouth shut. I had stopped fighting him many years earlier, when I went to college and finally escaped our stressful breakfast-table debate forums. Whatever my brother or I said at 7:00 a.m. over our bowls of cereal—it could have been an opinion about a homework assignment, an observation about the weather, or an expression of exhaustion—was met with a contradictory statement. Arguing was a sport for my father, an essential skill for success in his field. Therefore, it was a critical part of our upbringing. "You've got to get your juices flowing in the morning," he would say. "I'm training you. This is how you get ready for your day in the real world."

It got me an ulcer diagnosis during my freshman year in college. It also stopped me from sending any exploratory tendrils out into the "real" world.

"So, I need you to make me a flyer I can send to the other club members," my father called to announce one day. "I've got to raise thirty thousand dollars to set up an enclosure." The enclosure, he explained, would be a way to cordon off a chunk of the forest, ostensibly to see how the oaks did when Bambi and his parents weren't chomping away relentlessly at every sapling in the area. And it had an extra benefit: The American Chestnut Foundation people would sell him some of their blight-resistant hybrid chestnuts if he could get this thing built and ensure the young trees would be protected from deer.

"So, there's this stuff called ski fencing. Have you ever heard of it? It's this plastic mesh . . ." I cut him off. By that point, I'd lived in an Idaho ski town for twelve years and had a season pass to the local resort for most of them. I was quite familiar with ski fencing. "Well, anyway, it's ugly. Orange, with holes. But it's perfect, really, because you can move it around." I knew this too. "I've designed and priced an enclosure arrangement that involves ski fencing and concrete pilings. I think it can work." He went on to say that he envisioned barricading ten acres for seven years, then moving three of the four lengths of fencing to close off an adjacent ten acres for seven additional years, then repeating that pattern twice more. "So, we get twenty-eight years of data on how the forest does without deer, and we get to plant four crops of chestnut trees." His voice got louder. "In twenty-eight years, the chestnuts will be huge! Not that I'll live to see them, but someone will."

Of course, his optimism rests on the assumption that the chestnut tree's story will continue as projected—that all of these trees will take root, that the new genetic technology will enable them to fight off the blight, that some new pest won't disrupt their life cycle, that we humans haven't introduced another villainous plot twist into the ecosystem. In my worldview, that's too much uncertainty to maintain trust in my father's expected outcome. But that doesn't mean I'm not curious about how this all might pan out.

I made him a flyer. He sent it to his fellow club members. He did some heavy arm-twisting and, I suspect, threw down a chunk of the cash himself. Six months later, he called to tell me that he was on-site with the club's resident manager, watching him use a mini-cat to place the concrete pilings on the parcel my father had flagged earlier that month. "This is really cool," he said. "And I have twelve chestnut trees to plant. Anyway, gotta go. We're renting this machine by the hour."

He didn't ask me what I was up to. But I didn't really mind.

If you take a seed from a chestnut tree, plant it, water it, and protect it from predators, you can typically expect it to produce a chestnut tree that looks and acts more or less like the one you scavenged the seed from. Parents-to-be are often curious about how their offspring will uniquely combine their own genetics. Many people assume that their kids will be ideal mash-ups of both parents—in looks, habits, and interests. But it doesn't always work that way.

I suspect my father planned for me take over his company, although, like most complicated subjects in our family, this one was never discussed. He started bringing me into New York City with him, to his office, when I was about ten years old. My brother and I had fun playing with the intercoms, photocopier machines, and dot matrix printer paper, but I think my father was actually hoping to ease us into the Wall Street environment so that, when the time came, we would slide right onto the path he had prepared for us. He did everything right; he pushed the academics, gave me a taste for travel, sent me to the right schools. I got all of the sunlight and water I needed, and yet, I didn't blossom into the tree I think he wanted. Maybe I inherited an aberrant gene that gave me countercultural leanings, a wandering nature, and a lack of interest in material possessions and money. I often wonder if he would have used CRISPR technology to excise this gene at conception if he could have.

Shortly after I graduated from a stuffy East Coast college, I moved west and jumped from job to job for thirty years. Most of those jobs involved teaching in some regard—middle school children, yoga students, budding writers. Along the line, I spent thirteen years as a wilderness instructor, dragging suburban kids like the one I once was through Wyoming mountain ranges, Utah river canyons, and Patagonian fjords, hoping that the time they spent in the woods would foster an appreciation for the land and, of course, its trees. For a while there, I spent so much time camping for a living that I didn't bother maintaining a home. I didn't make enough money to be paying rent for a space I only occasionally slept in. Not only did I see places most people only glimpse in photographs; I developed relationships with them. I came to care as much about the land and its nonhuman inhabitants as I do about my own species. I now consider them to be my family.

I don't know what my father thinks of the story of my life. For years, I assumed that I was a source of disappointment, if not outright embarrassment. After all, he never asked how my students were, how my kayaking trip was, if I was dating anyone, or where I hoped to travel to that year. He just told me about the weather in New Jersey and reported on how his investment funds were performing, since we couldn't talk about the grandchildren he didn't have or the empire I wasn't building at work. And he routinely asked if I needed money.

On a recent to visit to New Jersey, I asked my father if we could go to the club to check out the chestnuts. He said yes and suggested that we spend the night there as well, since he also wanted to show me the trout habitat improvements the caretaker was working on. I packed my camera and journal and hopped into his hybrid SUV for two and a half hours of nearly conversation-free car time. I asked a few questions about the fund, about his cataracts, about the book he was reading and the Jeopardy champion he was following. The answers didn't offer openings for further exploration, so I mostly stared out at the deciduous forest flying by through the tinted windows.

After arriving at the club and driving out to the enclosure, though, he started talking. "Look at this. Look how different it is in here without the deer eating everything. Imagine what this forest would look like if they could be kept in check, if the red oaks could really get a foothold." He undid the clasp and pulled the orange ski fence over to the side. I could see six rings of metal mesh encircling green stakes and headed toward the closest one. "That one's doing well," my father said, as I approached a three-foot sapling with a few little branches and a dozen leaves. "And that one," he gestured to another tree, "is the go-getter of the bunch."

We walked around the plot, inspecting each of the saplings one by one. I took pictures of all of them, in part to document their progress for him and in part to look at later, to remind myself of what he saw in them—not just a handful of young living things but a handful of young living things he had invested significant effort into acquiring and nurturing. A handful of young living things he clearly cared about and had high hopes for. I knew, also, that he saw in them the future of a species in peril. I saw that too.

I paused in front of a rather sickly looking tree and asked if he knew what was going on with it. "I'm not sure. It's in the same soil as the others, and we dug the holes to the same depth." I bent over to look at its leaves. It had them; they were just smaller. "It'll do what it's going to do," he said, as we turned to walk back to the enclosure entrance.

I'm not as sure of my mutant status anymore. While I don't think my father understands my choices, it's possible that he sees some redeeming value in them, even though he wouldn't have made the same ones—even though they resulted in a daughter who fell far from the tree.

What I do know is that, whether he likes it or not, my choices have forced my father to find new ways for his influence to survive him. His parents lived well into their nineties, so the family genetics suggest that he could have five to ten more years on the planet. Still, although we never talk about it, I know he is thinking about what a legacy looks like without grandchildren in the picture. He recently told me that he set up a nonprofit arm for the club, one that allows him and other members to make tax-deductible donations to their conservation efforts. These include research into the effect of deer on red oak saplings, attempts to restore the stream banks, projects that help to naturally control the invasive gypsy moth population, and, of course, the ongoing cultivation of young American chestnuts.

I suspect some of the money he will end up donating to this nonprofit was originally sequestered in a college fund for the children I didn't have—an account that would have made sure these young humans would have the opportunity to thrive. It turns out the funds will go to making sure some of our nonhuman family members thrive. I'm more than okay with that.

If he cares about what I care about, then maybe, by associative property, he cares about me.

A couple of months ago, I was mountain biking with a friend along the spine of the Santa Cruz Mountains, about an hour from my home. I'd wanted to explore the trails off of Skyline Boulevard, the north-south road that traverses the crest of the range. I knew there were vast hillsides of California live oaks up there and that some of the trees had been planted in the 1700s. Seeing them would be well worth the trip. After planning a route that looked like it would maximize technical riding, Bay Area vistas, and oak spotting, we pedaled off to check out some new terrain.

About an hour into our ride, we arrived at a dirt parking lot, complete with an attendant and a few makeshift wooden stands. "Excuse me," I said, dismounting from my bike. "What's going on here?"

"We're a chestnut farm," the young woman replied. A chestnut farm in California? How had I not heard about this or seen it on the map? "We're open every weekend in October for u-pick. It's by reservation only. We fill up every weekend, pretty much."

In the process of firing dozens of questions at her, I learned that the farm's chestnut trees, like the oldest of the live oaks, were planted by the Spaniards. They're European chestnuts, not American chestnuts, but the nuts for sale in big mesh bags looked similar. She told us we couldn't enter the farm itself without a reservation, but we could skirt below it on a dirt road that had a good view up toward the trees.

I couldn't miss the chestnuts as we rode beneath the grove. They stood out among the twisty-trunked oaks with their straight, strong spines and big oval leaves. Their foliage was beginning to turn an autumnal yellow—something the majority of California wild tree species don't do. The chestnuts looked different from everything else around them, but they didn't exactly look out of place.

As we came around a bend, I spotted one chestnut tree right next to the road. A wayward nut must have rolled out of the orchard decades ago and sprouted a sapling outside the orchard's fences. I stopped and got off my bike. I had only ever seen full-grown chestnut trees in photos, I realized, and I wanted to touch one—one that, like me, had been transplanted to California to begin a new branch of its lineage's history.

I admired the tree's serrated leaves while I caught my breath. Then I reached up to grab a branch and pull it toward my face for a closer look. When I let go, the bough bounced right back up. I smiled, clipped back into my pedals, and kept riding.

Update: In the fall of 2023, the American Chestnut Foundation withdrew its support for the development and dissemination of the Darling 58 genetic variant. While the trees yielded promising results in labs and greenhouses, they did not perform as well as expected in the field. In addition, it came to light that an error had occurred in early stages of the genetic manipulation process. The American Chestnut Foundation acknowledged the public's existing fear of this technology, and its discontinuation of Darling 58 research is in part driven by its desire not to "erode public trust" in further research of this kind. It continues to pursue other promising paths toward restoration of the American chestnut to its historic range. For more information, see the foundation's website, www.tacf.org.

OWLGAZING

An owl allowed itself to be seen by me.

I was running on the other side of the nearly dry creek, on the trail that requires me to hide my sensitive skin under layers of protective fabric. I emerged from the poison oak–choked arroyo into the foggy fields of wild oat, fescue, and brome grass, then turned up the faint two-track, relieved to be out in the open. Just after veering onto the ledge with the ocean view, I swiveled my head toward the most prominent of the oaks and detected the kind of movement that can only come from a creature with a giant wingspan. I stopped and scanned the copse of gray and green tones, hoping to spot something out of the ordinary. I did—an oblong form, seemingly frozen onto a low-hanging branch.

It was a great horned owl, I think, based on its size and coloring. It looked at me—and I looked at it—for five dense and drawn-out seconds. We saw each other. And we were seen.

While I never thought I would be the kind of person who would marry young, I always assumed that at some point I would find someone to take on the world with. I never wanted kids, I wasn't looking for financial support, and I didn't think I needed to be seen as part of a couple. I just wanted a cockpit companion for shared adventures—things like music festivals, excursions to the hardware store, trips to South America, stressful family visits. Life.

After I'd been in a relationship with the mountain guide for five years, I started to assume we would eventually make our status official in the eyes of the law. We were in that late twenties to early-thirties age bracket, the period during which most happy-enough couples decide that the logistics of life are easier with rings. He and I had attended a series of weddings set in wildflower-strewn meadows, and the duos getting hitched were at the core of our social network. When I finally asked him if he saw us following in their footsteps, he said, "No. I've never even thought about getting married. Should I? Have you?" There was not a trace of anger, bitterness, or deception in his voice; the issue had simply never crossed his mind. Oily beads of shame oozed from every one of my pores, fueling the blaze I was witnessing—five years of perceived partnership going up in smoke.

A few months later, he got together with a coworker on a climbing trip. She was the one I'd once referenced in the midst of a fight, muttering, "Why don't you just hook up with her? She climbs and skis harder than me." They've been married for years now.

I probably should have learned something from this situation, but I was never really sure what the take-home was supposed to be. "Talk about your vision sooner" or "Have more frequent 'state of the union' check-ins" were the obvious ones. But it always seemed like there was something deeper to be mined, something I might catch a glimpse of if I caught it at just the right angle.

Last year, I attended a presentation at a local raptor center where the docent on duty introduced the crowd to a couple of truly majestic resident owls. "Let me be clear," she had said. "If you see an owl, it's allowing itself to be seen by you. I guarantee you that by the time you've spotted it, it's already been aware of your presence for several minutes." Owls can see in three dimensions and judge distances, much the way we can. They can also hear the heartbeat of their prey from three hundred feet away. "You will never sneak up on this creature," the woman said, nodding to the owl at rest on her gloved forearm, "although it may allow you to think you are doing just that."

Since then, I have made a point of recognizing my passive role in the act of seeing. If I am lucky enough to spot one of these elusive birds, I thank it for allowing me the privilege of its gaze.

I'm an almost fifty-year-old woman who has never been in a partnership that involved cohabitation. The mountain guide and I technically split rent at a few places, but he was working in the field most of the time, so really I was cohabitating with his dirty clothes and climbing ropes. After we split up, I invested a little too much energy into the kinds of men that single women semi-affectionately refer to as "bad boys."

I hung out with a guy who owned a garage door company and had more than a passing familiarity with the court system. He was an incredibly skilled skier and mountain biker and a hell of a lot of fun; he also said point blank on date number two that his ex-wife had been the woman for him and that there would never be another. I gave a few years to a charming half-Brazilian man who was quite open about his intention to find a woman in Rio de Janeiro who could supply him with a home and dual citizenship. Then there was a string of men for whom I served as a rebound girlfriend, happily and quite consciously providing some bizarre public service. I knew these relationships weren't going anywhere; my view of the world didn't line up with any of theirs. Still, dead-end dating somehow seemed preferable to getting blindsided.

My myopia is so severe that I am considered legally blind without my contact lenses. There is no question that, in another era, I would have been run over by a mastodon for the simple reason that I would not have been able to see it coming. My vision impediment first appeared in fourth grade, when I was called on to read aloud off of the blackboard and couldn't make out the sentence written in perfect pink cursive. I was the best reader in my class; I had whizzed through all the short-story modules in the color-coded box at the back of the room, and I was the only student who was permitted to dive into whatever chapter book she wanted during free reading time. But that day, during free reading time, I had to go to the nurse's office to squint at fuzzy capital letters hidden deep inside a big beige machine. In my fourth-grade school picture, I am wearing a pair of ugly wire-framed glasses.

We nearsighted people—those of us who cannot see objects from a distance—have what is called a "focal point." This is the dividing line between the part of the world we can delineate clearly and the part of the world that remains a total blur. Every year, my focal point moved closer and closer to me. By the time I was an adult and living on my own, it was only a few inches in from my face. Luckily, contact lens technology improved during the course of my childhood, and I have been sticking little plastic discs into my eyeballs for something like thirty-five years now. I put them in as soon as I get up every day and take them out right before I go to sleep, so I rarely experience my near blindness.

When I have almost cohabitated with men, I have deliberately gone to bed with my contact lenses still in my eyes. I take them out sometime later, when I get up to go to the bathroom. I tell my bedmate that I can't stand the awkwardness of groping for the glass of water on the nightstand. Really, though, I'm afraid I might miss some key facial expression.

My favorite bird at the raptor center was K2, a Eurasian eagle-owl—a slightly larger Old World version of a great horned owl. She was enormous, intimidating, and remarkably graceful, and my mouth dropped open as I watched her fly from perch to perch, swooping down with one beat of her wings to seize and devour frozen mice. Her feathers were covered with intricate weavings of rich brown tones, and her eyes were downright piercing. Apparently, she could kill a deer in the wild, and she regularly ate pocket gophers in a single gulp—bones and all.

I could see how developing some kind of a relationship with her would be emboldening. It's one thing to get a dog or a cat to be loyal; you just need to give them food and a little attention, and they're on your side forever. With raptors, it is clear that you need to earn their trust. You need to meet them where they are, play their game, and be on top of your own. I got the feeling that they would know when you were being dishonest—with yourself or with them—and that they would deny you their affection if they saw you entangled in a mess of self-deception.

Somewhere in the litany of bad boy companions, I detected and acknowledged my self-sabotaging pattern and started focusing on "nice boys." I figured they would want to know more about who I was and where I was going. I knew they might not be as much fun, but I had started to accept that someone with life-partner potential might present himself in some other kind of plumage than the kind I'd been on the lookout for. I decided to be more open about my interest in a long-term relationship—with myself and with them, and I was willing to overlook details that I had previously thought were important. Maybe it was okay if he didn't like to travel. So what if he didn't read much or invest political significance into every grocery store purchase? A little difference would be healthy, especially in the long view, which was what I was aiming for.

The nice boys supported me unquestionably in everything I did. And I did some really stupid things. I quit jobs and burned bridges and acted impulsively. I blamed other people for my bad choices, and I claimed victim status for a slew of misfortunes. They never called me on any of it. In their eyes, I was perfect. In mine, I was not.

Owls are classified as raptors. The word "raptor" comes from the Latin verb *rapere*, "to seize or take by force," so these birds are, in fact, defined by their predatory behaviors.

When I watch them, each and every movement they make strikes me as deliberate and calculated. These masters of efficiency appear to waste no energy on unnecessary action, and when they are still, they exude strength, confidence, and poise.

Sometime in my late twenties, the seemingly limitless deterioration of my eyesight finally tapered off. "Of course, soon enough, you'll start getting farsighted too," every eye doctor was quick to remind me from the age of about thirty on. For almost twenty years, I was able to come back at them with some snarky retort about my unique reverse-aging process.

When I had my annual vision checkup last week, I was confident I would be able to read the letters on the close-up card, and I was ready with a comment about being "frozen in my prime." But when the optician slid the farsighted test sheet in front of my eyes, I saw nothing but fuzzy, feathery, gray masses. Now I've got a pair of cute readers with hibiscus flowers on the sides. I've also got a reason to spring for another couple of sets with different color schemes. Then I can stash them in my desk and my car and my messenger bag, which is what I see my older friends do.

I'm not really all that worked up about aging. I like the gray skunk stripes emanating from my temples, and I appreciate the wisdom that I've gained along with them. It does worry me, however, that I now am not only unable to see things at a distance but am also unable to see what is going on right under my nose.

Supposedly, when one sense disappears, the others become heightened to pick up the slack. I wouldn't mind being able to hear the heartbeat of a mouse from three hundred feet away. Or the sound of my future sneaking up on me.

After I had cut things off with five attractive, kind, and relatively stable men over a six- or seven-year period, the word "predatory" began to be paired with my name in the landscape of small-town gossip.

I don't think my neighbors' assessment of me was accurate; no self-respecting predator would have wasted that much time and energy on such uncertain targets. I really believed that I could make these relationships work. At least, I did at the out-set. I tried to convince myself that people can grow into love, so I stayed with the bartender even after I realized that we would never have a conversation about the latest Booker Prize winner. I stuck it out with the ski patroller even though he complained incessantly about his job and threw back Jameson shots at the conclusion of each shift. I gave the groovy landscaper a couple of years more than I should have, hoping I would get better at navigating life with his toddler.

I didn't think I was gauging and measuring, and I sure wasn't trying to take anything by force; I was trying to slide into the kind of comfortable coupling I saw around me, the kind where each person can see the other one thinking, under-stand where they're coming from, and perhaps visualize where they might be heading.

The eyes of owls can constitute up to 5 percent of their body weight. *National Geographic* says that if we had eyes of the same proportion, they would be the size of oranges. Their enormous eyes make owls farsighted—like me, now.

The limitations of their farsightedness are overcome by their impressive depth perception and ability to see in low light. While they don't see in color—they lack sufficient cone cells to do so—their low-light vision is exceptional. Owls have five times as many rods in their eyes as we do, allowing for incredibly keen night vision. They see things we cannot. In conditions where we struggle to make out even basic shapes, much less fine detail, owls can both identify objects and locate them in time and space. It seems fair to assume that they can assess their significance too—in the black of night, from far away.

In each of my nice-guy relationships, I held on until eerily clear pictures of these men's futures began to emerge like individual trees from the forest of uncertainty. At some point, I started having recurring premonitions—like home movies, almost—of each of them going about lives that didn't and couldn't include me. They looked completely content in these dreamlike narratives, and it became impossible not to see how my presence would revise these scenes for the worse.

"You can't break up with me because you think 'you're not right for me,'" one said. "That's bullshit. I get to decide what's right for me, not you." But I stuck to my guns, convinced of my superior ability to see into the depth of time. I foresaw parenthood for several of them and a different sort of woman—more stable, more traditional, more rooted in routine—for them all. I clung to those images as I set each of them free while being accused of shortsighted thinking, restlessness, unrealistic expectations, and stupidity.

Two of them are fathers now. Three are happily (as far as I know) married. One is in a stable relationship with a single mom; he just moved closer to her in order to show his degree of commitment to both mother and daughter. And while it's true that one of the five is alone, I'm still quite certain he's going to make his way back to his ex-wife. Accurate timelines don't seem to be part of my limited clairvoyance.

The raptor center I visited rehabilitates injured birds and, whenever possible, releases them back into the wild. Many of the birds have been hit by cars; others have contracted lead poisoning from eating bullet-ridden carcasses, and still others are the victims of electrocutions or window strikes. After injured hawks, eagles, and owls are brought to the center, they spend anywhere from a week to several months undergoing treatments such as chelation for poisoning and surgery for repairing broken bones.

The last great horned owl to be admitted to the center came in just two weeks ago. He was hit by a car but suffered no fractures. They were able to release him the next day, after a course of anti-inflammatory pain medications. I know this because I check the raptor center's website obsessively, wondering what birds are in there and why. I like to know when they are set free, even though the center rarely reveals exactly where these former patients are flung headlong into the next stages of their lives. Sometimes they post videos of the moment of release. They're only four or five seconds long—even in slow motion—so I can watch them again and again if I want to feel that vicarious exhilaration. I don't need either of my pairs of glasses to see their lightness when they take to flight.

Were my visions real or were they just excuses? While they did play themselves out with intriguing accuracy, it could be argued that they were generic enough to be reasonable scenarios for guys in any small town—especially one in which the majority of residents sink their heels in and stick it out in their established nests. But if they were excuses, what were they excuses for?

"Let's just call this habit 'preemptive narration,'" a writing buddy told me over a couple of margaritas. "You script a story in which you can't possibly be right for the guy, which gives you an excuse to never really show yourself. You keep the plot line handy until you get restless; then you release him in some dramatic performance. When these guys end up living out something that vaguely resembles your little fiction, you get to be the hero, the catalyst, the fairy godmother. The one who enables him to spread his wings. Sounds to me like a great way to keep yourself hidden and still take credit for improving everyone's lives. Well, everyone's except for your own, that is."

I suddenly felt like my drink was a little stronger than I could handle, and my menu was looking extra blurry. Using an early-morning meeting as an excuse, I politely got up from the table and walked home.

There are about two hundred species of owls. Almost all of them are solitary. Up to eight owlets may be found in a nest together when they are young, but once they fledge, they are rarely seen in the company of others. Owls come together only for the couple of months of nesting season; the rest of the year, they live and hunt on their own. Great horned owls mate for life, however. Somehow, they manage to find and get back together with the same partner, year after year. I wonder if they recognize each other by their feather patterns or the rhythms of their hearts? Or perhaps it's something else: a distinct odor, a flight cadence, or even an aura that scientists can't yet detect. I wonder, too, how they found each other in the first place, how they established that they were a they.

I'm not sure what I don't want them to see. Once upon a time it was my imperfection, but I would like to think I'm over that. I'm pretty forthright about my confusion these days, and I issue plenty of caveats about my existential dilemmas. I fess up to my exercise addiction, my rigid eating and sleeping habits, and my inability to maintain a meditation practice despite knowing that it might be the only thing that can save me from myself. I readily admit that I've fenced off a minefield of unresolved parental issues, and I'm old enough to know that any physical perks I once offered are fading fast. What's left to hide?

I worry that by the time I am seen, there will be nothing left to see.

The owl and I both maintained our distance from each other. I knew better than to advance even one more step; just a twitch of my leg could prompt it to fly from the twisted limb into the darkness of the creek bed below. At the same time, those penetrating eyes appeared quite content to check me out from up above. For the owl, getting closer wouldn't reveal any new information. Its acutely refined senses had already taken in and processed everything important about me. On the other hand, I think I could have learned more about my companion if I had been able to see it better. Maybe it would have looked less regal, less powerful, or less smart if I could witness a twitch, shudder, or blink. Maybe I would have caught a glimpse of hesitation or ambivalence. But since I couldn't close the gap between us, I was left reinforcing my initial impression: that this bird was poised, wise, and in command of itself and its world. Everything about the owl's presence suggested that it could never be caught off guard.

The fog was lifting, and the fields of strawberries and brussels sprouts along the coastline were coming into focus. The faded map on the kiosk back at the parking lot had indicated that another path—one I had never been on—led down that way. I pulled my shirt off, tied it around my waist, and began running toward the ocean. Then I suddenly stopped and rotated my head to look back at the oak grove. I must have sensed its gaze, still unwavering, as it followed me along this new trail.

IBEX TROPHIES

I wanted to see an ibex badly, and I was not shy about telling everyone I knew.

"They're just goats, Bridget," several of my friends said. "What's the big deal?"

It's true; ibex are in the family Bovidae and the genus *Capra*, putting them solidly in the company of the farm animals whose milk I consume. But ibex are a special sort of goat: one who lives in the mountains. Big, steep, formidable mountains, like the Swiss Alps, where I was heading in August 2023 to do a challenging point-to-point trail run called the Via Valais. Completing the route was one of my goals, but so was seeing an ibex. I really didn't know why I wanted to see this particular animal so badly, but the idea had set its hooks in me years earlier and wasn't letting go.

My first glimpse of an ibex was in an armchair mountaineering book, probably one of those coffee-table tomes by photographer Galen Rowell that I often gawked at in my late teens and early twenties, when I was itching to climb remote, glaciated peaks. I remember a picture that captured the animal in silhouette, its body fully in shadow against a dramatic splash of alpenglow. What struck me most was the creature's truly distinctive feature: his horns. Not only are a male ibex's horns enormous, but they are also dramatically curved. They have ridges—bulging bracelets of keratin that manage to make the

horns look more like sculptures than body parts. Shortly after I saw that photo, I learned that, despite their unwieldy head-dresses, ibex are incredibly agile climbers. They regularly scamper up alpine slopes that would take human climbers hours to ascend. YouTube videos of ibex scaling the near-vertical, almost featureless faces of dams attest to this. As someone who was intent on shaping herself into the classic "mountain girl," a younger me found inspiration in these hardy creatures.

The ibex has a reputation for being elusive, and there's no question that this was also a big part of their draw. In addition, I wanted to witness firsthand the classic silhouette I had first seen decades earlier. Since that time, I had not only seen many more photographs of ibex navigating dramatic terrain, but I had also gotten used to spotting their likenesses on everything from Swiss beer logos and clothing labels to hotel signs and T-shirts. In fact, the ibex silhouette has been used so frequently in mountain town marketing that it's become something of a meme, a graphic shortcut to remind visitors of where they are. I wanted to strip away all of these branding connotations and see the animal for what it really was—in the environment it inhabited rather than on an aluminum can.

Most of all, however, I wanted to see a creature who had returned from the brink of extinction. I had read their come-back story, and it's a good one.

While alpine ibex had been widespread across the European continent since the Pleistocene, their numbers began to decline in the 1500s. In addition to being hunted for sport and for meat, they were harvested for a host of other body parts that our species' fertile imagination deemed valuable. A powder made from their horns supposedly cured impotence. Stones found in their digestive system, known as bezoars, made for lucky talismans. A cross-shaped bone from the heart of an older male ibex became an obvious pendant with which to adorn the human neck.

The advent of firearms didn't help the ibex, nor did their propensity for hanging out in groups in wide-open, treeless areas. They were easy targets. They yielded a lot of valuable resources and therefore fetched high prices. And the fact that they're found only in remote and rugged areas granted successful hunters a certain cachet. By the 1600s, ibex had been eradicated from Switzerland and Germany, and by the early 1800s, they had disappeared from Austria and Slovenia as well. In 1850, there were fewer than one hundred of the animals left in the Alps. All of them were living in the Gran Paradiso range, a stretch of mountains in northwestern Italy. King Victor Emmanuel II, the leader of that newly united country, was the kind of hunter who wanted to preserve the population for his own purposes. He also liked to impress foreign leaders with his nation's big-game options. So, in 1854, he created a Royal Hunting Reserve in the area and paid local poachers to work as de facto rangers, prohibiting anyone but himself and his guests from killing the remaining ibex. A half century later, the population was back up over three thousand, and Gran Paradiso had become a national park.

A Swiss hotelier named Robert Mader saw this success and asked the Italian king if he might purchase animals for a reintroduction effort in his country. He was denied. King Victor Emmanuel wanted a monopoly on alpine ibex—or, more accurately, on access to hunting them. They were not an endangered species to him; they were a political bargaining chip. Mader was not deterred, however, and in 1906, he paid Italian poachers (it's not clear if these were the same men being paid by the king to guard the animals) substantial sums of money to smuggle fifty-nine kids across the border into Switzerland.

Whether this effort was about preserving the species or reinvigorating a marketing effort is not clear. The flag of the canton of Graubunden—the alpine southeastern sector of Switzerland where villages like St. Moritz and Davos are located—features an ibex, as do the coats of arms of at least fifty Swiss villages. No doubt Mader and others knew that having at least a few of these iconic animals wandering in the mountains would make for good PR and increased local pride.

A captive breeding and release program facilitated the reintroduction of ibex into the Swiss National Park in Graubunden in 1920, and their numbers started to increase. When the Italians realized their herd was threatened by the animals' extremely limited gene pool, they did, finally, begin exporting them to other alpine countries. These efforts were assisted by the fact that, by this time, the gray wolf—the ibex's principal predator—had been all but exterminated from Europe. By the 1960s, about seven thousand ibex roamed the continent. Today, that number is over forty thousand animals. They're a severely inbred forty thousand animals, but they're out there and doing relatively well.

At a time in history when an estimated eighty-two species are going extinct every day, I find the ibex's rebound inspiring. On some level, I suspect that in looking for them, I was looking for hope. Perhaps if I saw an ibex, I could believe that we might slow this rate of extinction, that we could make room for at least some of our nonhuman companions.

Of course, the future of this species is not guaranteed. Alpine environments are among the ecosystems most severely threatened by climate change. Spring has arrived in the Alps a few days earlier each decade, shifting the life cycle of the plants on which the ibex survive in ways that don't line up with the animals' birthing calendars as well as they once did.

In addition, warmer temperatures are increasing the range of harmful parasites such as worms and ticks, pests that, in the past, hadn't affected ibex in their previously inhospitable homeland.

This inhospitable homeland is the same one that I regularly found myself daydreaming about in the months leading up to my trip. I had spent about ten days in Switzerland in the summer of 2022, and, despite having wandered in mountain ranges all over the world by that point in my life, the trip provided me with my first exposure to the Alps. From the instant I laced up my running shoes in Wengen, a tiny car-free village in the Bernese Oberland, I felt exhilarated. I loved the juxtaposition of jagged, craggy peaks and rolling green pastureland. I loved the lushness of the larch forests and the preponderance of berries, mushrooms, and wildflowers tucked into them. And I loved the settlements I ran among, with their hand-hewn log cabins bedecked with geranium-filled window boxes. In the two weeks I was there, I found myself running longer and longer distances, simply because I wanted to spend as much time as possible bathing in the nourishment of this landscape. No part of me wanted to get on that plane back to the United States. By the time it touched down on my home soil, I was already plotting my return.

That plotting began to include the idea of a point-to-point trail run—the kind of adventure that involves moving from town to town and hut to hut with a minimal amount of gear and without ever doubling back or retracing steps. I considered a number of different routes before I settled—with a fair amount of trepidation—on one called the Via Valais. My hesitance was justified; each day of the journey would involve between twelve and twenty miles of running with between two thousand and seven thousand feet of elevation gain. Not Stairmaster-style elevation gain, mind you, but rocky, uneven climbing at altitude—the kind of thing ibex excel at but humans need to train for. I considered easier options, but the Via Valais was in the Valais, the canton where the famous mountain towns of Zermatt, Evolène, Verbiers, and Saas-Fee were located, the canton with the really big mountains. So I committed to the route, but I made sure a few rest days were built into it. As a fifty-three-year-old runner, I figured my knees would thank me for a bit of time off along the way. I was still nervous, but I knew that most of my days had bailout options. If worst came to worst, I figured, I could avoid parts of the route by taking a train or a bus around to the next town and meeting my luggage there. That wouldn't be "completing the route," but it would at least keep me on the move and seeing new landscapes.

I spent the summer running uphill with the pack I would be wearing in the Alps, figuring out what foods I could best digest when exercising for eight hours at a time, and I prayed that my tendinitis wouldn't act up. Then, in late August, I flew to Geneva with one giant duffle bag and two big goals.

The technical term for the curve of the ibex horn is "falciform," which is generally translated as "sickle-shaped." I've seen the latter phrase, along with "saber-shaped" and "scimitar-shaped" used on hunting websites, where you might expect such weapon-oriented language. To me, ibex horns look like oversized, perfectly curved commas hanging in midair, suggesting the pause that the sight of this majestic creature warrants. But of course, I'm a writer. Regardless of how you perceive these horns, you can't miss their annual rings, bulges of material about a half-inch wide that traverse the front surface. The width of these rings is affected by factors like diet, weather, and herd size, and the number of rings allows you to determine the age of the animal—if you can get close enough to count them. For the most part, the only people who have that opportunity are those who have shot them.

I cannot begin to imagine deliberately killing this graceful creature who has come back from the brink of extinction. Yet it is done. These days, most ibex hunting is "trophy hunting," the shooting of animals for the express purpose of taking a body part—a head, a skin, a set of horns—as a souvenir that affords bragging rights to its owner. Most of what I learned about ibex horns came from a web page dedicated to the evaluation of these kinds of "trophies." It was filled with intensely detailed sentences like, "An important factor in assessing the overall quality of the trophy is the relationship between horn length and horn mass over the whole length (horn mass index)." Here we are, humans ranking the value of animals' body parts, just like we rank the value of the creatures themselves. Until recently, the canton of Valais sold about a hundred hunting permits per year, grossing upward of $650,000 in the process. These hunters just took the heads and horns. Since they were not allowed to fly the meat out in a helicopter, the bodies were left to rot or be eaten by other species.

Of course, if it weren't for hunting, the king of Italy would not have set aside land to preserve the ibex. Then again, if it weren't for hunting, the ibex probably would not have become endangered in the first place.

This is not the only complexity related to the comeback of the ibex. Between the Italian king's desire to keep them as potential hunting trophies for his royal cronies and the Swiss theft of the kids, the whole process strikes me as ethically murky. When it comes to morality, I don't tend to fall into the Machiavellian camp, the one in which the ends justify the means. But here we are, with forty thousand ibex in Europe. I am grateful for that.

My gratitude also feels a bit complicated by the fact that I've found the best information about ibex on hunting websites. The people who kill these animals seem to be best acquainted with their habits. Besides scientists, they're the ones who have really bothered to get to know them.

I've never killed a mammal. I've never really stalked one either—certainly not for days on end, learning their habits, dietary preferences, and diurnal schedules like my elk- and deer-hunting friends have. I typically see animals in passing, while I am on my way to somewhere else. I believe that I value them for themselves, not for what they do for me or my species. But my valuation seems more theoretical. And my intimate knowledge pales in comparison with that of most hunters.

On August 28, I arrived in the hamlet of Arolla (population two hundred), the starting point for the Via Valais. The sky was a vibrant late-summer blue as I got off the Swiss Post bus and humped my duffle bag up to the hotel. Rain was in the forecast, however, and it was predicted to begin the following morning. I wasn't scheduled to start the official route the next day; I was supposed to do an "acclimatization" loop from the hotel that would get me into Les Aiguilles Rouges, Arolla's well-known, picturesque pinnacles. When I learned that my room wasn't yet ready, I made a quick decision to do the acclimatization run that afternoon so as to not waste a good weather window.

I was glad I did so when, the next day, I ended up walking down the town hill to check out Arolla's four hotels, two restaurants, and three stores in a downpour. One of those stores was a gear shop, and while I didn't need any more expensive techy mountain garb, a T-shirt in the window caught my eye.

It had the word "Arolla" written across the top in a very 1970s font—one I remember from a much smaller T-shirt an eight-year-old me was wearing (along with a pair of roller skates) in a 1978 family photograph. As befitting that vibe, the shirt in the Arolla window had a rainbow on it, and featured inside this rainbow was the silhouette of an ibex. Of course. I might not be much of a shopper, but I still get excited when I see an article of clothing that has the potential to become a lifelong companion. "They probably only have it in men's styles," I told myself, but I went into the shop anyway. Five minutes later, I walked out with a perfectly fitting women's medium, in a great shade of faded Levi's blue. "I might not get any farther than Arolla on this trip," I reasoned, given the way the weather was looking, "and I might not see an ibex either. But I can celebrate my pursuit of both with a shirt."

When I got back to my hotel, I checked MeteoSuisse, the website that offers town-specific weather predictions, for the seventh time that day. I had heard that there was a very real possibility of snow in the mountains, and while I wasn't lacking warm clothes, running shoes were the only footwear I had. The forecast looked quite unsettled, and I was supposed to travel up to the Becs de Boisson mountain hut at ninety-eight hundred feet of elevation the next day.

When I woke up to misty but not totally darkened skies and saw no snowflake icons on the MeteoSuisse page for at least ten hours, I felt good about heading down valley toward Evolène and starting the long climb from there up to the hut. My selfies from that day show me first smiling in front of a dripping wet barn with a hood pulled tight around my face, then posing in the midst of a herd of Hérens cattle in just a T-shirt.

I got to the Becs de Bosson hut at about four in the afternoon—tired, hungry, and wet from both sweat and intermittent rain but grinning from ear to ear. I was in the Alps.

I had completed one day of the Via Valais. After collapsing onto a bench in the hut's foyer, I pulled off my saturated running shoes and picked out a pair of size 39 plastic Crocs, the required indoor footwear in Swiss mountain huts. Before slipping them on, I gave my tired feet and tender calves a quick rub and thanked them for getting me through the day.

While I waited for the hutmaster to serve me and the other four guests a heap of rösti, a classic Swiss potato pancake dinner, I scanned the bookshelves above the wood-burning stove for intriguing reading. In addition to lots of climbing magazines from the late 1990s and early 2000s, there was a glossy coffee-table book dedicated to ibex—or, more accurately, to *boquetin*, since the text was all in French. I leafed through it with wide eyes. Page after page showed males perched on ridgelines and groups of females with their much shorter, lighter horns lifting their heads from the hard work of foraging as if to admire the scenery from their dinner table. There were photos of ibex scaling cliff faces, scampering across scree fields with their young, and trudging through knee-deep snow with ease. It was like a journal of life in the mountains, for a creature who has evolved perfectly to dwell there.

I tried to dwell in the mountains at one point in my life—"dwell" in the way that human beings can, which is to say, by carrying food, fuel, a portable stove, and a lightweight shelter. I spent thirteen years working for an outdoor leadership school, leading groups of eight to fifteen students on wilderness excursions that included thirty uninterrupted days in roadless, uninhabited mountain ranges. By stringing a few of those contracts together and minimizing the days in town between them, I could almost pretend that I was actually living in the mountains. But when we met horsepackers for a resupply every ten days, it became clear that we were only temporary residents, fully reliant on outside support. The staff refilled our fuel bottles, gave us new bags of food, and took our trash away. They even passed along emergency messages if necessary and dropped our outgoing letters in the mail when they got back to town. Despite having spent upward of 160 weeks' worth of nights in a sleeping bag, I can confidently say that I wouldn't last more than a couple of days above tree line without the equipment and rations I need to carry in from the outside world.

So these days, I no longer try all that hard to do so. I spend a lot of time in the mountains biking, running, hiking, and skiing, but I emerge from the woods in the evenings and sleep in a bed. I do trips like the Via Valais, during which I am out in the elements for long periods of time but end the day in front of a smoldering fire with a roof over my head, eating something other than pasta.

My inability to survive independently in an ecosystem I love contributes to my respect for ibex. They have adaptations I do not. For starters, their diet is perfectly suited to their surroundings. They mostly eat grasses, mosses, and flowers during the short alpine summers; during the other months of the year, they forage on shrubs, evergreen needles, and lichens. It's a low-calorie diet, so they need to spend much of their days grazing, but it's a reliable one. And they store substantial fat reserves that they can metabolize during the winter when food is scarce. They have thick, long wool that keeps them warm all year-round, and their lungs are larger than those of other animals of comparable size, giving them an aerobic advantage at higher altitudes. In the locomotion department, the muscular legs of the ibex help them climb in steep terrain and are powerful enough to enable jumps of up to six feet. Their hooves are made of a spongy material covered with a thin layer of keratin, a combination that enables their feet to mold and adhere to a cliff face while also protecting them from cuts and scrapes. Like all arteriodactyls—hoofed mammals with two toes—they can move those two toes independently, creating the kind of stable platform that's needed to perform their impressive rock-scampering feats. Finally, they've been known to hunker down in caves or tunnel into the snow when the inevitable storms hit.

I know all of this from research, not from days—or even hours—of observation. Does that make this knowledge less valuable, less significant?

I wondered at times if I "deserved" to see an ibex, if I had done the kind of work that warranted an interaction. Maybe all I had really done was purchase a plane ticket—an action that certainly wasn't helping their species or any other.

On the second day of my Via Valais run, I woke up, put on my headlamp, and walked down the stairs to the hut's bathroom. Along the way, I glanced out the large dining room window. It appeared lighter than it should have for the hour—because everything was white. I opened the front door and was blasted with a gust of wind. I looked around and estimated the snowfall to be about a foot. It was still August.

The five of us who had passed the night in the hut spent the next couple of hours looking at maps, looking at weather sites, and looking out the window. Two of the other guests were running the same route I was, so we decided that, at the very least, we would band together as we considered our options. We could wait out the storm in the hut for a day or more, we could follow the footprints of the hutmaster who was planning to hike out to his vehicle that day (though his vehicle was nowhere near where we were heading), or we could stick to our intended route. Of course, we couldn't see our intended route, since the only visible features outside were the snow-covered picnic tables on the hut's front deck. But the couple running my route had a GPX file—a digital map of the route—on one of their phones, and though we didn't have any service, we decided that we could follow the dot as it moved along the big blue line. Our plan was to make our way to the reservoir below us and then try to get on a road that would eventually lead to the town of Zinal, our intended destination.

In the time it took us to make that decision, another six inches of snow had fallen. Another foot would fall before we made it to that reservoir, four long, cold hours later. Our cheeks stung and our fingers—stuffed in only light liner gloves—were numb. It was often hard to see each other from fifty feet away, leading to a persistent sensation of vertigo. Nevertheless, we couldn't stop moving for fear of becoming hypothermic.

When we were about to run down the cell phone's battery, we finally took a break to regroup under the eaves of a shepherd's shelter. It was then that we spotted the paved road that ran along the dam. It, too, was covered with snow, but the heat of the asphalt had caused it to melt in a way that distinguished it from its surroundings. Had there been anyone else out there, they would have heard our whoops and hollers through the thickness of the cloud layer and seen my shoulders descend inches.

After walking the road for about twenty minutes, we saw a little Renault sedan and flagged it down. My mediocre French managed to communicate our status: cold, tired, and still many miles away from Zinal. The nice couple who were, unbelievably, just going for a joyride to visit the dammed lake in the midst of a late August blizzard, drove us about six miles down the road to a point from which we could run another paved road to Zinal.

As we approached Zinal just before dusk, we started seeing a lot of the dark brown, wood-sided cabins characteristic of villages in the Valais. Some of them displayed an assortment of skulls. They were of mixed sizes, and I assumed that the smaller ones were chamois—a shorter, darker-furred, lower-altitude-dwelling cousin of the ibex. But I knew some of them were ibex skulls; there was no mistaking the giant ridged and curved horns.

Even though these eerie skeletal ornaments were mini-versions of hunters' trophies, it felt like they were welcoming me to the village after a very long and stressful day. It also felt like they were telling me to be grateful for what I got. We had veered pretty far off-route just to make it to Zinal, and it was impossible to say what the next few days were going to bring. My goal of completing the Via Valais as written—with all of its twists and turns and mountain passes—was not going to be achieved. And the likelihood of my seeing ibex in the snow was close to nil. But I was running with two new friends toward a hot shower, a change of clothes, and a big fat dinner. I *was* grateful.

I must admit that I sometimes feel a sense of superiority when it comes to wildlife interactions. Not that I'll cop to it often, but I tend to think that I am a little more stealthy than other people—and a little more sensitive, too, especially when I'm alone. I like to think that I am more apt to notice a noise, a glimmer of fur, or even just the presence of another animal. In all likelihood, this is groundless hubris, but I think it also helps me see wildlife by functioning in the way that self-fulfilling prophecies do. Because I'm convinced that I have a special ability to spot creatures in the wild, I work to prove myself correct. I scan the landscape constantly, attune my ears to the sounds around me, and stay as quiet as I can. I even talk to the creatures I can't see but hope are nearby, figuring that putting out a vibe that communicates my respectful intentions can't possibly hurt. Then, when I do see one of my nonhuman companions on the trail, my opinion of my talents is reinforced in what scientists call a positive feedback loop. This might all be nonsense, but it's working for me.

It didn't, however, work for me over the course of the next couple of stages of the Via Valais, during which the only living things I saw were trees and a handful of other trekkers. Everything and everyone else were under snow.

After a rest day in Zinal—one largely spent in my hotel room hiding from the whiteout and nursing my stiff butt and back muscles—I had two days of difficult travel, first to the Turtmann Hut, then to a town just south of Zermatt called St. Niklaus. While the sun came out from time to time, it did very little to melt the several feet of accumulation that had fallen up high. I postholed—the term mountain travelers use to describe punching through snow up to your shins, knees, or thighs—for hours up and over passes. Luckily, I was walking in at least one set of footprints at all times, so I did not have to search for the trail or follow a blue dot on a cell phone screen. But I sure wasn't running, even though I was in running shoes lined with plastic bags in an attempt to keep my feet semidry. I wasn't seeing any ibex either.

As I neared St. Niklaus four days after the storm, the sun started to come out, and I slowly emerged into grassy slopes and pastures. The last two hours of that run had me traversing the Höhenweg trail, cruising along the western side of the Mattertal Valley at sixty-nine hundred feet of elevation through a mixture of larch forest and clearings. As I removed my jacket and once again exposed my arms to the light, I saw mushrooms of all sizes and colors, raspberries that had sprung back up after shedding the weight of the snow, and mosses glowing with new injections of moisture. I didn't see any animals, but I was so happy to be running on solid ground through greenery that it didn't matter. I stopped in a tiny town called Jungu at a café stationed at the top of the world's tiniest gondola. As I wolfed down a slice of lemon cake, I stared at the peaks on the other side of the valley. There were some famous ones there, like the Seetlhorn, the Platthorn, and the Nadelhorn, but they were still enshrouded in clouds. I took some pictures anyway, then ran downhill to my hotel and my flip-flops.

I was scheduled to have a layover day in St. Niklaus. However, when it turned out to be a perfectly clear and cloudless one, I abandoned the idea of "rest" in favor of riding the world's smallest gondola back up to Jungu. At first, I wasn't sure if this was such a good idea, since my knees were starting to announce their displeasure with the amount of downhill pounding they were being forced to endure. I wondered if my choice might jeopardize my completion of the rest of the route. But I ultimately decided that if I rode the lift up and down, I wouldn't really be taxing my body. I would be able to see the summits of the peaks I'd only glimpsed the bases of the day before and sample the other item on the café's menu: apple kuchen. In addition, I'd noticed a trail that left from the café, heading in the direction opposite the one I had arrived on. I figured I could go for a walk or maybe an easy run on it. I packed my running vest with a few layers, some water, and a credit card and made my way to the gondola.

Naturally, once I got to Jungu, I decided that I needed to earn my apple kuchen by getting at least a little bit of exercise. I popped into the café to make sure they still had a few slices left—then I turned onto a trail called the Wildblumenpfad. Every quarter mile or so, there was a sign posted with a photo of a wildflower and its name in German and Latin—like "Alpenblumenweg Jungun, *Carlina acaulis*." I wasn't sure the

German names would stick in my mind, but I recognized some of the Latin ones from other mountain ranges, since some of the same alpine flowers exist on multiple continents. I kept on going after the signs petered out because it became apparent that there was a wide valley above me, about a mile ahead. I popped up onto that shelf, poked around the little grazing cabin next to the stream, then found a sunny rock up above it. I sat and ate the two rolls I had taken from the hotel's breakfast buffet along with a slice of cheese and a mini packet of Nutella. Then I headed back down off of the shelf and toward the wild-flower trail.

I was crossing a boulder field about a half mile before the wildflower ID signs started up again when I heard the sound of rocks shifting. It was sharp and sudden, and it snapped me to attention. I froze in place. Immediately, I stopped trying to remember the handful of new German words I had learned that morning. My mind became a blank slate, ready to take in and process whatever happened next.

It was not my feet that caused the sound, and there were no other human beings around. I scanned the grassy slopes above me and saw nothing. Then I heard the rocks shift again, and this time, I could place the origin of the sound: ahead of me, to my left, and up above the trail. I squinted in that direction and held my breath. Then I spotted something: a stocky, dark brown set of haunches. A moment later, I watched a head swing to look downhill. In the process, two giant comma-shaped horns appeared, standing out against the light-colored rock behind them. I was looking at a male ibex, seemingly alone.

"Hello," I said softly, as much to myself as to him. "I'm just going to stay here and watch you for a little bit, okay? You just keep doing your thing." And he did. He bowed his heavy head back down to the earth, his horns curving forward like the bottom rails of a rocking chair in the process. I imagined that he was focused on the freshly watered plants eking out a living between the rocks, finding them especially juicy and nourishing.

When I got over the thrill of spotting this ostensibly lone male, I remembered that ibex rarely travel by themselves and scanned the area upslope of him. Sure enough, there were two more males, about fifty yards uphill, walking slowly toward the ridgeline. As I tracked their progress, I saw three more animals, two of them females—or juveniles—both with their heads down, eating. And then three more, this time a little higher up and back toward the ledge I had come from. I couldn't make out their horns or even their legs; I just saw their bodies and slow movements.

Nine ibex. I saw nine members of an incredibly elusive species. And I saw them on a day I wasn't even supposed to be running around in the mountains, a day when I wasn't sure if my body could carry the weight of my ambitions. Witnessing the ibex gave me a shot of energy, and I ended up running back down to town rather than taking the gondola. If they could keep going, so could I.

When I Googled "trophy hunting for ibex," I was over-whelmed by the results. You can travel to twelve different countries to shoot one of these creatures. One site I perused offered a three-day trip (with "a near 100% success rate") for fifteen thousand euros. That did include meals and lodging but not the "species harvest fee" (twenty-five hundred euros per animal), taxidermy costs, or alcoholic beverages. On the site's home page, below all of these details is a picture of three middle-aged white men posing with a dead male ibex. The man in the center is smiling and holding up one of the horns so that its head is lifted up off the ground—its normal position in life.

But the creature's eyes are closed, and his legs are splayed out at an awkward angle beneath him. I cannot read his expression in death.

Looking at this photo makes me nauseous. However, despite my gut reaction to ibex hunting, it is not currently a threat to the species. The canton of Valais—the only area of Switzerland in which trophy hunting by foreign nationals was allowed—decided to end the practice in 2021 as a result of pressure from citizens. Valais residents can hunt ibex, but stringent governmental regulations have ensured that the process will not affect population numbers. Some of the resources I consulted claimed that the number of ibex taken in the yearly hunt is roughly the same as the number that would have to be culled to maintain a healthy herd.

Some experts believe that, these days, the biggest threat to ibex survival is outdoor recreation. In other words, *I* am the threat.

To run the Via Valais, I flew across the Atlantic and took two trains and a bus to Arolla. I paid a company to drive my luggage from hotel to hotel. I stayed in mountain huts built with the assistance of helicopters and stocked by them as well. The use of fossil fuels for all of these activities is affecting the climate in ways that are detrimental to the survival of the ibex. On top of that, my presence in the terrain might be making the situation worse. This thought makes me almost as nauseous as the sight of the hunter holding the horns of the dead male.

How different am I, really, from a trophy hunter? Was my goal of seeing an ibex just another version of a trophy? Was I somehow taking a piece of them too?

I did finish the Via Valais—although because of the wild snow day, I can't say I truly completed the route. Not that I care. After I saw the ibex, I had two more days of running around Zermatt. The trails were dry, the air was warm, and the skies were cloud-free. I cried when I first saw the wildly dramatic profile of the Matterhorn, Zermatt's famous needlelike peak, and then I cried three or four more times, just because I was so overwhelmed by beauty. I was then—and still am—terrified that I won't be able to hold on to that beauty, to store it, to call it up later when I get depressed by the state of our planet and my role in setting us on the path we seem destined to travel. At the end of the route, I ran into the lobby of my Zermatt hotel ten pounds lighter than when I started but infinitely more grateful for my body and the access it affords me to the natural world. I think about that wild stew of emotions, including both the guilt and the gratitude, every day.

I also think about my ibex sighting every day. I wear my 1970s-style Arolla/ibex T-shirt once a week, and on those days I think about it even more. It felt magical to me in the moment, and that magic is lingering. Having had contact—however brief and distant—with a creature who thrives in an environment I love and yet can't quite exist in touched me in a way that reading about ibex and seeing photos of them never did. The experienced changed me, though I am still struggling to put into words exactly how.

I struggle, as well, to understand whether or not this change is important enough to justify the resources that went into making it happen. The carbon footprint of a transatlantic flight is enormous. To relieve myself of my guilt about it, I left my home county only twice in the two months following the trip, putting only one tank of gas in my car during that time. In addition to avoiding mammal and bird meat, as I have for decades, I ate no seafood and no cow dairy, baked my holiday gifts instead of buying them, and purchased nothing other than food for at least a month. Does any of this count? Is it worth maintaining these personal balance sheets when oil companies still drive national agendas and agribusinesses still clear enormous swaths of forestland to meet our species' demand for beef? I would love to say that I'm not involved in these large-scale destructive activities, but of course I am. I heat my home. I use fossil-fuel-driven transportation. It's complicated and messy, much like the ibex's comeback from the brink of extinction. There are too many factors involved, and the math just gets fuzzy.

I did try to take a few pictures of the ibex on my phone, mostly so I could study them later. Perhaps I could count the rings on their horns or see the color gradations on their fur, I thought. And, yes, I wanted both bragging rights and some proof that I had stumbled my way into one of my goals. Not surprisingly, I deleted most of the photos. The creatures were just too tiny and too well camouflaged to see amid the rocky slopes, and zooming in on them only yielded pixelated blobs. Still, I kept a couple to remind myself of the encounter and help me later recall the terrain in which it had taken place. I'm the only person who can see an actual animal in these images. They, too, are fuzzy. What I'm really seeing is my memory of seeing the animal—which is something altogether different. And the question remains: Is that memory, and the empowerment it inspires in me, significant enough to justify the steps taken to acquire it, or is it just my version of a trophy?

Maybe what I do with that "trophy" going forward will determine its ultimate value.

For the last week, I've been working on a woodblock carving of an ibex to use for my New Year's cards—the ones I use both to connect with people I care about and to hopefully inspire them to think about the fate of the nonhuman creatures with whom we share the planet. The design is shaping up to be the classic male-with-long-horns-silhouetted-on-a-ridge image—the kind that gets used for beer labels and Swiss coffeehouse logos. Only, for me, the ibex doesn't have those connotations anymore. Instead, he's carrying the weight of multiple narratives: the story of his species' near extinction and amazing comeback; the stories of the hunters, poachers, and hoteliers who drove them to that point and then brought them back; the story of my ibex sighting and the run that precipitated it; and the as-yet-unwritten story of how these creatures will fare as their alpine environment warms—as well as my role in that process.

That's a lot of weight to carry on uneven ground. It's a good thing they've evolved to do just that.

ON SANDERLINGS, THE COLLECTIVE, AND ME

I've heard people call sanderlings "the energizer bunny of shore-birds." It only takes a few minutes of watching them to see why. They are perpetually in motion—first chasing, then running from waves in an endless flurry of activity. Their skinny little legs move so quickly that they blur. When those legs stop for a few seconds, their beaks take over, working like jackhammers to excavate invertebrates from the wet sand. It's rare to see a sanderling standing still.

It's equally rare to see a solo sanderling. Unlike their larger shorebird companions—sandpipers and plovers, for example—sanderlings roam the beach in packs. The group I stopped to observe today on Main Beach in Santa Cruz, California, had twelve birds in it, all running at full speed toward the retreating edge of the water. When they got there, they stopped and dug like mad for two or three seconds before a new, weaker wave surged forward. In unison, the birds shifted just enough to avoid getting their feet wet. As soon as that water began to recede, heads popped up and stilt-like legs took off in pursuit of the freshest treasure. Then the ocean lapped up once again, and once again the sanderlings ran for safety. This dance continued for as long as I had the patience to watch it. Dash forward, forward, forward. Dig, dig, dig. Dash back, back, back.

Pause for a momentary regrouping—then forward, forward, forward, all over again.

What amazes me about these little birds, in addition to their apparently boundless energy, is their remarkable ability to move as one. When they chase a wave, they *all* chase the wave, executing nearly identical movements while maintaining nearly identical spacing between individuals. When they retreat from a wave, they are in lockstep. In addition, they seem to know how the ocean's pulsation will play out. It's as if they can antic-ipate exactly where and how the foam will pull back from the sand, making their skittering movements as much a part of the wave as the wave itself. Or perhaps it's more accurate to say that they are their own wave, propelled by something like a group consciousness.

As a twenty-first-century American, I don't have a lot of experience with group consciousness. I don't even identify myself as a member of a group, much less as someone who runs, hunts, and eats within the boundaries of one. I think this is why I can't stop watching sanderlings. They make me, someone who has always defined herself as an individual alone in the world, wonder what it would be like to truly exist as a part of a whole.

When I was in sixth grade, I was chosen to be part of my middle school's "Gifted and Talented" program. Even at the age of eleven, I thought both this title and the concept behind it reeked of snobbery, but I didn't have the guts to say so. Instead, I happily attended our special class sessions because I got to miss a few periods of social studies, my least favorite subject. Doing logic puzzles with the only kids around who didn't tease me for being a "brainiac" was definitely more fun than talking about Colonial Williamsburg for the third year in a row.

Our district's Gifted and Talented program was geared toward participating in a statewide problem-solving competition called Olympics of the Mind. This was New Jersey in the Reagan years, so everything—including creative enrichment activity—had to have some head-to-head showdown that resulted in a ranking. We trained to beat the neighboring towns. I can't remember exactly what we needed to beat them in, though I vaguely recall building some kind of a battery-powered cart that could collect random objects scattered about in a maze.

What I do remember with great clarity is that these endeavors promoted teamwork. In retrospect, I suppose they were meant to prepare us for collaboration with other software designers or medical device engineers in our productive future lives. At the time, I was puzzled by them. "Why can't I just work on this myself?" I asked the faculty adviser. "And what's going to happen when I do all the work, and Jamie just goofs off the whole time? Are we going to get evaluated on our group performance?" Nothing about my academic career up until this point had entailed working with other people. I studied on my own, answered questions on my own, and succeeded on my own. I just didn't see why I needed to be matched up with other kids to do anything.

I don't blame myself for having this attitude at age eleven, although I'm ashamed of it now. I was a product of my upbringing, and my upbringing emphasized individual distinction and achievement above all else. I was praised for having the clearest handwriting, for getting the highest scores on standardized tests, and for having perfect attendance. Even in group activities, like team sports, success was somehow twisted into how well I did—how many batters I struck out or how many goals I scored—regardless of whether or not my team won.

It's no wonder I begged my teachers to let me do all "group work" alone. They almost always let me. Still, I was curious. Was there something to this idea of collective action—something to be gained by throwing my lot in with other members of my species?

We humans are accustomed to observing sanderling-style group awareness among airborne birds; many of us have witnessed flocks of starlings, for example, mid-murmuration, as thousands of winged creatures swoop, dive, and bank turns as if tethered together by filaments of gossamer. But we rarely see this kind of movement on the ground, and we rarely see it right in front of us. The uniqueness of this combination is what forces me to stop and watch sanderlings at every opportunity.

Because the birds are so accessible and indifferent to my presence, I can't resist pulling out my phone to film them. I see how close I can get without startling them, then stop and zoom in—enough to allow me to see each bird but not so much as to miss out on the collective movement of the group. Then I press record and hope I can keep up with the flow.

I bet I've taken hundreds of these videos. While I'm waiting for a prescription refill or killing time on an airplane before takeoff, I watch my mini-clips on endless repeat. I scroll in on one bird, then the next, then the next, trying to determine which of the sanderlings initiated the shift in action. There must be someone in charge, I figure. But it doesn't look like there is, no matter how closely I study the unfolding action.

Not long after being recruited into the Gifted and Talented program, I discovered MTV. Since I was able to do my homework and watch videos at the same time, my grades didn't falter even as I spent more and more of my time memorizing Hall and Oates lyrics and learning the names, ages, and hometowns of all five members of Journey.

MTV was a gateway drug into the wonder that was 1980s arena rock. I was lucky enough to have a mother who was willing to indulge my burgeoning addiction—probably in part thanks to the fact that school was still going well—by driving me down the highway to stadium shows at the Meadowlands, often with a couple of girlfriends in tow.

The way I see it, Duran Duran's 1984 *Seven and the Ragged Tiger* tour performance at Brendan Byrne Arena in East Rutherford, New Jersey, was my first experience of something like group consciousness. Sure, I had sat through hundreds of Catholic church ceremonies that were supposed to enable me to experience transcendence, but where they failed, Simon, Nick, John, Andy, and Roger hit it out of the park. I had never experienced the single-minded focus of so many people. For starters, every person there (except for maybe my mother and a few other chaperones) knew every word to every song and chanted the lyrics in perfect time with the band. In addition, when a song really dug into its groove, people jumped up and down and stamped their feet along with the beat of the kick drum. This made the floor—the concrete, rebar-reinforced sports arena floor—move. The sheer collective enthusiasm of twenty thousand concertgoers literally shook the foundations of the building. I didn't know that could happen. When the band did a call-and-response thing, the audience shouted what they were told to shout. When the band asked how the crowd was doing, the audience screeched in unison. When the band slowed down for a ballad, the audience yanked out their Bic lighters, held them aloft, and swayed.

I don't remember what I was wearing that night, what homework assignments I was worrying about, or what I might

have said to any of the friends who were there with me. In
retrospect, I can't even remember which songs the band played,
though I bet I can guess. What I do remember is being awed—
not by the band's talent but by the crowd. The individuals in
attendance had become one unified being. I also remember
that, although I was singing, dancing, and jumping up and
down, I felt like I was losing track of my body and its boundaries. And I know I liked it—so much that I went to as many
more Brendan Byrne Arena shows as I could before I left for
college.

I wondered, too, what it would have been like to be a
member of the band. Did they experience a similar sensation of
melting into the collective? For them, was the collective their
five-man collaboration or did it extend out to the audience?
Either way, I could see how they were doing something as a
group that couldn't be achieved as an individual alone. That
struck me as both scary and cool.

I always assumed that the packs of sanderlings I watched
were some sort of socially cohesive group, like an extended
family or an interbreeding network. I knew that most other
creatures who hunt, run, and eat as units are somehow knit
together. They form gaggles, harems, tribes, and other kin-
ship-based structures and develop working relationships based
on these ties. It seemed obvious to me that, in order to be able
to move so fluidly, sanderlings would have to be familiar with
the other members of their groups and have a great deal of
experience reacting to their movements—like the five guys in
Duran Duran.

I only recently learned that this is not the case. Sanderlings
don't show what scientists call "flock cohesion"—a fancy word
for stable group membership. They don't stick with some set

roster of birds when they are running onshore or flying. When groups come together and then split, the new subgroups that form typically contain different individuals.

At first, this was baffling to me. How could they possibly participate in this seamless and graceful dance with total strangers? Were there rules to be followed? If so, were they instinctual or learned? If not, was the ability to flow with the flock just part of their nature? Or was there actually some kind of collective consciousness at work here—either one created by the coming together of individuals or one that always existed, that the individuals tapped into?

Regardless of how their harmonious movements happen, sanderlings are more like teenagers at arena rock shows than I had originally thought. Both are syncing up with creatures they've never even met.

About ten years ago, my mother and I traveled to Turkey. In addition to wanting to see the Hagia Sofia and eat as much baba ganouj as possible, I was determined to attend a ritual performance of the Sufi sect known as the Mevlevi—or, as I had often heard them called as a kid, "whirling dervishes."

I was doing a lot of yoga at the time, and I had become intrigued by the idea of losing one's self in some kind of spiritual trance. Despite hundreds of hours of sweaty, crowded workshops and attempts at seated meditation, I never got all that close. While I remained hopelessly stuck in my own brain and body, however, I was fascinated by people who claimed to escape theirs.

I had read about Sufis who spin themselves into oneness with their creator, and although I was not itching to try their technique, I had spent enough time making myself dizzy over the years to picture how repetitive cyclical movement might precipitate an altered mind state. I was eager to see what this practice looked like, and my mother—the same fearless woman who was willing to drive teenage girls to arena shows—was game. We did a little research to determine which *sema* (dance ceremony) was most authentic and bought our tickets.

Soon after we took our seats in the circular, wooden-floored space, the music started—first with a steady droning undercurrent. Then, layer by layer, other instruments and voices added additional flavor to the sonic stew. I call it a "stew" because there really was no detectable melody. My Western-trained mind searched for it, looking, as it always does, for something distinct and individual to grasp on to. Finding none, I resorted to letting the tones wash over me.

The men, known as *semazens*, entered single file and stood in a line for the length of a song, giving me the opportunity to notice their differing heights. Although they all wore identical beige stovepipe hats (called *sikke*), I noticed that they each had slightly different facial hair, or no facial hair at all. They were, in fact, different people. After a few minutes, they removed their black shawls to reveal their characteristic ritual outfits: stark white shirts, jackets, pants, and the billowing skirts that, once the men begin spinning, form roughly six-foot-diameter circles around their bodies. One by one, the *semazens* stepped forward, bowed, and began a sequence of precise steps. With their left feet anchored, they pressed off the balls of their right feet to "whirl" in a counterclockwise direction. They first crossed their arms in front of their chests, then slowly extended them toward the sky, holding them at shoulder level like wings, extensions of their hearts that helped them soar.

Once all twenty men were in motion, I lost the meager ability I had to tell them apart. They held their bodies in the same posture, used the same foot sequences, tilted their heads at the same angle, and spun at the same rate. As is the case with sanderlings, the spacing between the whirlers' bodies remained perfectly even throughout the ceremony. Despite having their eyes half-closed the whole time, no one tripped or fell, and no one broke rank. For the next twenty minutes, they became a unified mass and movement—a rotating, undulating corona of white.

I couldn't help being affected by the beauty of their motion. And I had questions: Did the same men perform every night? How well did they know each other? Did they have to practice in order to avoid colliding, or did they just naturally feel each other's presence? Would they ever get together after a ceremony to talk about what they had experienced? And what *had* they experienced?

It is said that the Mevlevi whirl to know God. The *sema* is often referred to as a form of prayer, an act of remembrance, and a representation of witnessing. While the stamping of feet supposedly symbolizes the ongoing process of eradicating the ego, it seems to me that dance itself catalyzes that very process.

While I was observing the ceremony, I would first zoom my vision out, taking in the overall pattern of the dance; then I would zoom in and watch one individual. Out, in, out, in. This pattern, along with the repetitive, groundless music, had me nearly falling asleep several times. Only the jerk of my head falling toward my shoulder brought me back to the room. It wasn't that I was bored or tired; it was more like my boundaries were dissolving. I was fully immersed.

Yet, when it was over, I was also quite pleased to step outside into the cool air of the April evening, zip up my jacket, and find the edges of my own unique body.

A group of sanderlings is called a "grain." What an odd collective noun. It makes me think of rice, millet, or wheat—foods I eat when they have been cooked to death and made into a relative mush. I rarely pay attention to individual grains. They fall unnoticed onto the floor or slide down my throat as part of a spoonful of calories.

When I watch sanderlings, I notice their every movement, although, admittedly, I have a hard time telling them apart. Like many birds, sanderlings have juvenile plumage, breeding plumage, and nonbreeding plumage. Since I live in their winter range, I am most used to seeing white-bellied individuals with a gray-and-white-speckled backs—nonbreeding adults. I see some juveniles as well, who typically have more black on their backs, sometimes forming what birders call a checkerboard pattern. Some of the dark plumage extends from the tops of their heads toward their eyes, like a widow's peak. The two plumage patterns are different, but not radically so. Breeding sanderlings add a few additional colors to their palette, though. "Rust-colored wash" and "rusty mottling" are two of the phrases I have seen used to describe this distinctive seasonal outfit.

Of course, they could be spray-painted pink, and I would still recognize them by their behavior—their amazing ability to move as part of a group.

Shortly after my MTV heyday, I left the public school I'd grown up attending and went to a private school where everyone was new to me. I quickly discovered that, at this institution, each of the cafeteria tables was strictly identified with one particular group of kids. There was an artsy table, a music geek table, several jock tables (the lacrosse and field hockey girls formed one, and the soccer and basketball girls formed another), and a misfit table. I'm not sure now if I identified them on the basis of their plumage or their behavior; I think it was a little of both.

For the first couple of weeks of my freshman year, I tried out a few tables. I wanted to sit at the edge of each, testing it out without being fully identified with its label. But the tables were circular; there were no edges to hover around.

After a month or two, I gave up altogether and sat in the side room, where the staff members and friendless kids ate their lunches. I scarfed down my food in five minutes and then went to the library. That seemed preferable to having to run with a group for four solid years. Yes, I was alone, but I at least I was an individual and not just another indistinguishable glasses-wearing girl at the nerd table.

And yet, I kept begging my mother to take me to more arena rock shows.

I also started spending a lot of my weekends taking the commuter train into New York City in order to ride subways around Lower Manhattan. I rode in the most crowded cars I could find and always stood, clinging to the cracked vinyl hand straps while glancing at the variety of faces around me. I got jostled just enough to know I was moving but not so much as to invade the personal space of my neighbors. There seemed to be rules about how close you could get to another rider, and nearly everyone adhered to them, constantly making minute adjustments to their space cushions. As the wheels screeched around corners, everyone swayed in unison. When the train stopped, everyone absorbed the energy of the car by bringing their torsos back to center. And after the doors opened, we swam like a school of fish into the urban ecosystem.

Out on Sixth Avenue or West Houston Street, I would join another current of humanity, being careful to walk at the same pace as the crowd. Sometimes I had a destination in mind—an art house movie theater or one of the dozens of thrift stores in Greenwich Village—but other times I was just going with the flow. I was completely anonymous, yet I was not a bystander; I was a moving piece in the machinery of a city on the go.

Sanderlings are capable of migrating enormous distances. The majority of them breed in the Arctic regions of Canada and Alaska, where marine invertebrates, their primary food, are plentiful. This allows the migrating adults to eat well, but, more important, it allows their newly hatched young easy access to a feast. After breeding, sanderlings typically head south for the winter.

While the migratory destinations of some species (bar-tailed godwits and sandhill cranes, for example) are quite specific, sanderlings end up in a wide variety of locations. Some birds will migrate only one thousand miles to southern Alaska or coastal Canada, while others will travel up to six thousand miles to South America. Sanderlings can be found on almost any tropical or temperate beach in the world, reminding us that they have a lot of options to choose from when deciding where to overwinter. And they appear to have the ability to make that choice for themselves.

That choice can even be not migrating at all. Some birds— nonbreeders, in particular—determine that the long trip up north is not worth it and choose to stick it out in, for example, coastal California, where I live. I see them year-round on my local beach, so I've got to assume our fertile waters, dramatic tidal fluctuations, and relatively temperate weather are appealing enough that they see no reason to go elsewhere, even if the majority of their companions opt to make the trek.

Whether they stay or go, they obviously find other sanderlings with whom to do their feeding dance. The more I watch it, the more I wonder what it feels like to them. Do they crave it? After a while, do they need space away from it? Do they, like me, feel the tension between wanting to do their own thing, be their own thing, think their own thing and wanting to join a collective that seems to create a larger consciousness—one in which they can lose awareness of their own individual struggles? Maybe it's not a tension but rather a perfect balance. I certainly wish that for them.

Or maybe there's no need for balance because, for sanderlings, it's possible that there was never any individual/group division in the first place. There's a good chance it's only my species that wrestles with these questions, that it's only my species that overcomplicates the routine activities of life on Earth.

Many years after my Duran Duran phase, I graduated into the jam band scene—one in which I still roam. Most people will attribute the founding of this musical lineage to the Grateful Dead, the band that pioneered the idea of improvisationally oriented (and pharmaceutically assisted) rock and roll concerts. As any Deadhead will tell you, the band's thirty-plus years of live shows were quintessential experiences of collective creativity. "We used to think of ourselves as being fingers on a hand," guitarist Bob Weir says in the documentary *Long Strange Trip*. "They can move independently, but . . . they're all connected at the core."

Each of the band members has spoken about the individuals' willingness to lose themselves and their attachments to structure and outcome onstage, claiming that this release and its resulting fluidity fostered the group consciousness responsible for their most magical moments. In addition, every band member has gone on record saying that their audiences were always a part of that consciousness, that the energy and intentions of every person in the stadium contributed to the new and innovative versions of the songs that resulted night after night. "I remember distinctly receiving literal communications from the audience," says bassist Phil Lesh in the same movie. "Nothing in words. Sometimes it was actual musical ideas, little fragments of melody." Keep in mind, this is a band that initially practiced syncing up by regularly dropping colossal amounts of acid—a chemical known to facilitate ego loss—prior to picking up their instruments.

It's probably a good thing I didn't discover the Grateful Dead when I was in high school. In retrospect, they were offering what I was looking for. But I might have gotten lost while finding it—lost in the drugs, lost in the flow, lost in the crowd. As an adult with an arguably overdeveloped sense of individuality, however, I am fascinated by them. I follow a whole host of Grateful Dead spin-off bands, Grateful Dead cover bands, and what's come to be known as "Grateful Dead-adjacent" bands. I read articles that analyze their lyrics, listen to archived recordings of shows, and lap up documentaries about the band members.

I love the idea that my presence at a concert makes the music better, that my participation makes everyone's experience richer. It makes me feel powerful. It's not the kind of power people allude to when they talk about CEOs and politicians, of course; that kind doesn't interest me in the slightest. It's the kind of power that comes from being part of something much, much larger than yourself.

Apparently, I'm still looking for that—complicated or not.

I am completely baffled by religious traditions that believe an individual lives on after death. And there are a lot of them. It's not just the Catholics I was raised with who think we still get to have our personality quirks after we pass from this plane; most Christians, Jews, and Muslims agree. No matter how hard I try to wrap my head around this idea, I just can't buy it. I imagine this is why the experience of group consciousness

I'm looking for can't be found in institutional religion. I believe that when I breathe my last breath, there's no more "me." Full stop.

Somehow, this conclusion manages to be both totally terrifying and incredibly freeing. When I'm not clinging to the idea that my individual self is the pinnacle of creation and deserves immortality, I actually find great solace in the idea that I will someday return to the energetic whole. I like to consider how my body will be dismantled and reduced to the elements from which it was formed. I used to think it would be fire or fungus that would do the job, but I recently read about a company that performs burials at sea. I've grown to like the idea of being dumped on the floor of the ocean where the protozoa who form the base of the marine food chain—along with the mighty force of water itself—will be responsible for my decomposition.

At that point, molecules of carbon or nitrogen or oxygen that once made up my calf muscle, left ventricle, or pinky fingernail will become part of something else—perhaps a diatom or a stingray or an octopus. Or a tiny crustacean who gets gobbled up by a sanderling on the run from a surging wave. At that point, I will have totally fused with whatever universal collective consciousness might exist out there. There will be no "I" to eat doughnuts, run mountain trails, or boogie down to the Grateful Dead's "Eyes of the World."

In the meantime, however, I'm still me. But I'm becoming a more permeable me—a me who has come to see the value of the collective, a me who might be able to participate in that collective without dismantling the container of the individual. A me who, like a sanderling, can flow between states.

Recently, I read that sanderlings have only three toes, unlike most other birds, who have four—three in front and one in back to help them keep their balance. Apparently, sanderlings lack that rear-facing toe, technically referred to as a "hallux." This keeps them leaning forward into the future, and it also helps them run really fast toward and away from the waves they chase.

I had never noticed this before, so I went down the beach this evening with the objective of looking more closely at their feet. But between the wet sand, the water, and the birds' incessant leg movement, I can't say I got a good glimpse of their anatomy. I also wasn't able to stay focused; as usual, I got sucked into the big picture. I spent my time watching the wave of white, brown, and beige feathers as it advanced and retreated with the water, marveling, as always, at its rhythmic pulsation.

Although my gaze was fixed on the sand, I noticed a gray mass out of the corner of my eye and raised my head to look out toward the horizon. Flying parallel to the shore were hundreds—maybe thousands—of dark gray birds. They were sooty shearwaters, gull-like creatures known for their aerial acrobatics, epic migrations, and enormous flocks. Although they are fairly common central coast visitors, their prodigious numbers and impressive synchrony never fail to command my attention. I watched them as they all swooped down closer to the water, then suddenly pulled up and reset their trajectory a bit farther up from the surface. They all jogged left, toward me, then back right toward the open ocean, rising and falling together on currents of air I couldn't see.

I watched them until they disappeared behind the wharf; then I went back to observing my sanderling friends, the other manifestation of collective consciousness happening right in front of me. Even though I witness their feeding dance almost every day, I couldn't resist taking another video. It might help me remember that what I seek is all around me, as long as I open my eyes to it.

ECHINODERM ENVY

I wasn't looking for a sign, although I'm never averse to receiving one. Even if I were, a sea star sighting on New Year's morning would still be memorable. I had been anxious to put the passing year to bed, so I crawled under the sheets early in hopes of accelerating the process. At dawn, a beam of coral-colored light nudged me awake and onto Seabright Beach, where the high tide had wiped the sand clean.

My auspicious echinoderm sat just a few hundred yards from the lighthouse, on the sand just above the wrack line, where it was displayed like a king's crown atop a bed of moist purple kelp. It was a big specimen—about the length of my bare foot—with a flesh-colored body. The dark mauve edges of its five limbs faded to a paler pink shade along the axis of each arm. From above, it looked a lot like a little man. Its top appendage resembled a regally extended head, the two flanking it celebratory arms—one with an upturned hand, the other a downturned palm, like a statue of the dancing Shiva, the Hindu deity of destruction and rebirth. The sea star's remaining two legs looked like human ones, their tips curled toward each other in a cute pigeon-toed posture.

I shook my head as I caught myself wantonly anthropomorphizing—again. No matter how hard I try to avoid it, my brain always seems to be working to make other creatures more like me. But sea stars aren't like me at all. For starters, they're

invertebrates, so they lack spines—the backbone type of spine, that is. Some species do have the other kind, the prickly kind that extend out from their skin into the world where we can step on them. When we humans classify animals, our first dividing line is "spinal cord/no spinal cord." We've ranked the "spinal cord" category (vertebrates) solidly above the "no spinal cord" one (invertebrates) on the ladder of living things, even though 90 percent of animals on the planet fall into the backbone-less group. Our rankings have to do with our ideas of complexity and evolutionary advancement, not popularity or success. Or the possession of a superpower—like the sea star's enviable ability to jettison a part of itself and start over again with a brand-new one.

I would happily take that skill on any day of the year.

When I was young, my father introduced me to the term "analysis paralysis." He teased me for my inability to make even the simplest decisions, such as what to order from the six-page Chinese takeout menu or which of my closetful of sweaters to wear to school. "Life is a series of choices," he said. "You have to just make them, move on, and never look back."

That seems to be easier for some people than others—people like my father, an investor who makes hundreds of risky financial decisions every day. Not people like me, who read those 1980s *Choose Your Own Adventure* books twenty-plus times in order to experience all of their possible endings: "Turn to page 28 if the princess enters the forest; turn to page 34 if she backtracks to the cave." I would turn to page 28, wander through the woods with my princess protagonist, and run out that story; then, an hour later, I would go back and see what happened when she poked her nose into the cave. I spent weeks with these books, tracking which trajectories resulted in happily-ever-after and which got the princess killed by a bloodthirsty wolf.

My analysis paralysis persists to this day. Not only do I have trouble making choices to begin with, but I question them for hours, days, or years afterward, looking back obsessively at the big ones in an attempt to understand how they culminated in the tiny details of my present reality. If my present reality isn't perfect, I figure, then there must be a decision that I should have made differently years ago.

When I walk Seabright Beach, I often find myself wondering what would happen if I could turn back a chapter or two and choose another path.

Nearly everything I've read about sea stars has included some variation on the sentence "Sea stars sever their limbs in response to danger." The lack of an infinitive verb like "decide to" or "choose to" in this recurring phrase seems significant. Since sea stars don't have brains or spinal cords, many people are reluctant to give them anything like agency in this process. The word for this capacity to drop parts is "autotomy"—a word that, interestingly, differs from the word "autonomy" by only one letter. I like to think there's some active decision-making happening, although it may look and feel very different from ours. But, then again, I've already confessed to my tendency toward anthropomorphism.

When one of the sea star's limbs treads into dangerous terrain—perhaps a young sea turtle has gotten an arm into its mouth, or a sea snail has started to chomp on some of the creature's tube feet—a group of neurons registers that risk. *It's time to make the break*, they say. Then another group of neurons starts the severing process by triggering a reaction in what's called the "catch connective tissue." This tissue is found only in echinoderms, and it varies in stiffness depending on which proteins are activated in the body. When a sea star is exercising its autotomy superpower, its catch connective tissue becomes so soft that it essentially dissolves. The part of the animal that very recently joined the main body with the arm ceases to exist. Like magic, the individual cuts ties with a part of itself.

Does a sea star feel anything in that process? Is there physical pain from the severing of flesh? Is there emotional pain from the loss of a limb? Or is there some kind of relief, like the feeling of starting over on New Year's morning?

I always thought I would stumble upon the perfect career path, even though I never had an inkling of what it might be. All throughout high school and college, my parents and guidance counselors kept telling me not to worry about my lack of clear direction. "You'll figure it out," they all said. "Just keep exposing yourself to a wide variety of opportunities."

I did that. In my twenties, I worked for the Forest Service, for my father's lawyer, for a music management company, a tutoring service, a speed-reading outfit, two public schools, a raft company, and a handful of other places that I can no longer remember. After my thirties and forties passed by without yielding some kind of progressive employment pattern, I came to the conclusion that I just never tried the right jobs.

I can imagine myself in any number of dead-end situations from my past, activating my catch connective tissue to call a do-over. In one, I am a twenty-four-year-old, applying for teaching credential programs, sitting in classrooms and attempting to convince myself that I like young children more than I actually do. Sometime between that exploration and the time I finish up another degree and start applying for teaching jobs, my catch connective tissue starts tingling, telling me that I'm in a less-than-ideal environment. It makes the break before I get sucked into a miserable three-year detour through the bureaucracy of the California public school system. I'm saved the pain of disappointing my principal, my coworkers, and my parents, and I get three years of my life back to try out a career path that might last.

In another, I am thirty, working in environmental education and thinking about going to law school to become an advocate for the wildlands I live in and love. My boyfriend won't leave fieldwork and won't move to a city, so I put it off for a year, then two, then three. By the time he ends our relationship, I've lost the motivation for a long run of schooling. If I had catch connective tissue, it might have sensed the perilous nature of making choices on shifting sands. I might have gently dissolved that relationship arm long before he hacked it off, leaving me free to grow into something else.

Autotomy sounds pretty good to me.

When it comes to growing new body parts, most sea star species practice what is called "unidirectional regeneration"—meaning a body can grow a new arm, but an arm can't grow a new body. The detached arm dies because it does not contain part of the central disk, the nerve ring that runs around the sea star's mouth and contains its digestive system. The limbless main organism still has part or all of that disk, so it's capable of growing a new arm. That intuitively makes sense.

But there are a few wildly gifted species in which the abandoned appendage can grow back the rest of the organism, even without containing a piece of that all-important central disk. The scientific term for this process is "disk-independent bidirectional regeneration." I just call it mind-blowing. Sea stars from the *Linckia* genus—found primarily in Indo-Pacific waters and made famous by internet photographs of its electric-blue member, *Linckia laevigata*—have evolved this capacity. The arm of a blue sea star, which is skinny and long like a human finger, can live off of its own stored nutrients until it can grow back a mouth. Which is to say, it can grow a whole new body without even eating.

This is so foreign to my human idea of integrity—both physical and egoic—that I have to watch YouTube videos of regenerating individuals to even begin to wrap my head around it. It makes me wonder how much of ourselves we can get rid of before we cease to be ourselves.

What would it be like if all the lives I rejected—ones I've tried and given up on, ones I thought about but decided not to pursue, ones I turned my nose up at without giving them a second thought—regrew a new me? What would that me look like, sound like, think like? Probably not like the me that's sitting here today. She would be a product of different experiences, and I suspect those would make her a different person.

But would she be entirely different? Or is there some blueprint of her out there, some Platonic ideal or woo-woo soul essence that endures?

I hope for her sake she would be a little easier on herself.

You would think that, given their incredible regenerative powers, sea stars would be taking over marine ecosystems everywhere. Not so. In a number of places, including where I live on the central coast of California, sea stars have suffered from a mysterious and horrible disease called "sea star wasting syndrome," or SSWS. I've seen photo sequences of a sunflower star afflicted with SSWS that document the individual's literal disappearance. In the first photo sits a striking purple and orange creature, about two and a half feet in diameter with twenty-four limbs radiating from its center. In the next photo, taken a day later, several whitish lesions have formed on the sea star's arms. In the shot taken three days after that, all that's left of the animal are a few fragments of pale flesh isolated on a jagged rock and a gooey substance leaking from the remains.

It's not clear what's causing SSWS or why it was particularly devastating in 2013 and 2014 but has been less dramatic in the last few years. While a virus may be directly responsible for this condition, it's clear that warming sea temperatures play an important role in its increased prevalence. We know what's caused those.

The wasting disease photos eat at me from the inside. They make me wish I had chosen a career path that would have enabled me to save the sunflower sea star from this horrible fate. I could have been a marine biologist, like I dreamed of doing when I was young, before I realized that I would have to take upper-level math classes in addition to going scuba diving, observing organisms, and wanting to save them from my own species.

Instead, I'm here helplessly watching the beautiful creatures disappear before my eyes as I write about them.

Apparently, when a sea star first comes down with SSWS, it stops eating. Then it exhibits behavior some scientists describe as "listlessness"—a very human term for their lack of interest in food and movement. Their use of this word makes me wonder what the individual actually feels. Some people argue that sea stars don't "feel" anything at all, that they are too far down the evolutionary ladder for that. Once again, I'm not so sure about this dismissal, given that we're talking about creatures who can regenerate themselves using processes we don't yet understand.

I also wonder if their version of listlessness is anything at all like what I experience when my brain convinces me that not only have I made bad choices, but my choices have done noth-ing to help the world. When those voices take over, I lie down on the floor—on my back, with my limbs spread out, looking a lot like the little Shiva sea star—trying to take control of my breath. Sometimes I cry salty, viscous tears, but more often I just melt into the carpet, like I, too, have a wasting disease.

I don't though, so I always get up off the floor.

Not only can I not contract SSWS as a human, but I also can't separate myself from my past and grow a whole new future. And really, the more I stare at those disintegrating sunflower star photos, the more I think that might be okay. For years, I've been convinced that severing myself from my so-called bad choices would fix me. Without their overbearing presence in my consciousness, I've always assumed that I would experience some kind of blissful sense of freedom. But now, as I grimace at this sight of a stunning sea creature disappearing in a camera's eye, I wonder if I would just disintegrate without them. My choices— good or bad—might be what's holding me together.

I also wonder if my New Year's Day echinoderm encounter might be some sort of invitation. Sea stars, like nearly all living things these days, live in a human-driven *Choose Your Own Adventure* narrative. We could turn to page 28 to establish additional sea star nurseries, to page 37 to set aside more marine protected areas, and to page 66 to vastly reduce our carbon emissions. Or we could turn to page 79 to keep going about our business and allow them to disappear into the fossil record. I suppose I, too, can turn to page 12 to thrash around in the sea of my old choices for the umpteenth time, or turn to page 41 to read, observe, and learn more about this creature I get to interact with—the unique evolutionary wonder that is the sea star. Then I could sit down and start typing.

I suspect the closest thing I'll ever have to a superpower is the ability to share my awe of the myriad crawling, flying, swimming, leaping, replicating, and sometimes even regenerating creatures that exist around us. I could use it to help all of us regrow new limbs that recognize the importance of this awe. Or, better yet, arms that are willing to protect it.

After my New Year's walk to the beach, I sat down at my computer to identify my intertidal zone discovery. Its common name is the fairly bland "pink sea star." Its Latin name is a bit more intriguing: *Pisaster brevispinus*. In this case, the short spines being referenced are the little pokey ones on its exterior, not the ones that separate its classification status from my mine. I remember bending down to touch them and finding them to be surprisingly soft. I was expecting these little protuberances to react to me, perhaps because I know I would experience a whole host of sensations if a stranger touched my skin. But nothing happened. Right—they're still, in many ways, not like me.

I scrolled through my photos. I took fifteen in all, after standing on the beach for a long time admiring how the first rays of sun gave my companion dimension and how the foamy edges of waves attempted to reclaim it. One picture would have sufficed for identification; the other fourteen were to help me remember that I got to see a sea star on the first day of the year—a day when I needed a little extra help regenerating narratives.

In my favorite picture, there's a strand of bull kelp in the lower right-hand corner, a broken mussel shell in the lower left, part of a fish skeleton off to the side, and a big old human footprint at the top. It's not the most aesthetically appealing or well-balanced image in the group, but I still like it. I like the fact that the sea star isn't alone. We beach-loving creatures are a diverse gang, it tells me—one that's actually done rather well at sharing this earth/sea interface for hundreds of thousands of years. And we all have superpowers we can employ to keep that going.

I exported the photo and made it the backdrop image for my computer's desktop. Every day, when I sit down to write my part of the story, I am forced to remember that I'm not alone either—even if my company looks nothing like me.

SPONGES, THE ANTIHUMANS

When I rolled off the dive boat at a site called "Jager Bomb" in Roatan, Honduras, I hadn't been in the ocean with a tank on my back for about twenty-five years. Yes, I had done a scuba skills update a month before this trip, but it was held in the Boys' and Girls' Club pool a mile from my home in Santa Cruz, California. The water was ten feet deep, and the tile-covered wall was never more than a couple of kicks away.

Admittedly, I was nervous about my return to this activity, but I knew that I had loved it once upon a time, when I dreamed of swimming through coral caverns and meeting a green turtle's gaze for weeks after coming home from a dive trip. So I was back on a boat in the Caribbean Sea, listening carefully as Daniel, the divemaster—the paid guide with the advanced certification and knowledge of Roatan's reef sites—did his briefing. He said that we would descend to a water depth of about sixty-five feet and asked everyone to stay within a few body lengths of him. He flashed through a quick review of the industry-standard signals he would use for communication, such as a hand slicing across the throat for "out of air," a thumbs up for "go to the surface," and two fingers moving forward for "follow me." Then he whipped through a flurry of gestures used to point out underwater life. He closed his hand by touching his thumb to his fingers and said, "This one's for an eel." Grouper, lobster, shrimp, shark, octopus, ray, and turtle were presented so rapidly that they blurred together. "Got 'em all?" I did not; turtle (hands together, palms down, and thumbs out and waving) was the only one I remembered. But I assumed that, as long as they weren't emergency signals, I could just look to where he pointed and figure it out. Finally, Daniel told us we would be down for about forty-five minutes and clapped his hands. "Bueno. Let's get in there!"

While diving isn't a very athletic or performance-oriented sport, mistakes can be lethal. We humans can't breathe underwater on our own; it's only through the use of highly engineered equipment that we can hang out in the aquatic environment for more than a minute or two. Assembling that equipment properly is, therefore, a survival skill—one that was a bit rusty for me at that point. So just before I donned my whole rig, I did one more gear check. My eighty-cubic-foot aluminum oxygen tank was standing upright along the side of the boat with my BCD—buoyancy compensation device— attached to it. This critical piece of gear is an inflatable vest that allows you to wear your air cylinder on your back. It also helps you control your depth by changing the amount of gas in its built-in bladders. My BCD looked fine, so I shifted my attention to my regulator assembly, which was attached to the top of the tank, looking like the messy combination of black tubing and assorted devices that it is. I checked its four components, each of which hung from a section of rubber hose: the primary stage, the mouthpiece I would breathe from; the secondary stage, another mouthpiece in case the primary one failed; the pressure gauge that would tell me how much air was in my tank; and the coupling that attached to the BCD and enabled air to flow into and out of the vest on demand.

Earlier, I had twisted the big black knob to open the tank's valve, but I grabbed it again, just to make sure. I then proceeded to test all four parts of my regulator assembly one more time. I stuck the primary stage in my mouth, inhaled

the pressurized air, and exhaled through the regulator. Check. I did the same for the secondary stage. Check. The pressure gauge read 3,000 psi (pounds per square inch)—the standard amount for a full tank. Finally, I grabbed my BCD inflation hose, pressed one button, and watched the vest expand. I paused for a second, then pressed the other button to watch it deflate. My gear was good to go, even if I may not have been. I kept reminding myself that innumerable wondrous creatures were living their lives right below me and that, in a matter of minutes, I would have the chance to coexist with them.

I was already in my wetsuit, and my mask was around my neck, making the next item to put on my weight belt, a thick strap threaded through ten pounds of lead. The human body—especially one wearing neoprene—can be quite buoyant, so this piece of gear helps keep you underwater, especially as your tank gets lighter as a result of air consumption during the course of the dive. As I slung the belt around my waist, I remembered that it's always put on so that the right hand is used to release the buckle in the event that the diver—or the diver's buddy, if the diver goes unconscious—needs to ditch the pounds and get to the surface immediately. I hoped I would never even have to think about this, much less experience it. But in case I did, I took off the belt and turned it around since I had put in on the wrong way. I looked up to see if anyone noticed. Fortunately, the other divers were all deep into their own equipment navigation. Wondering what other details like this I might have forgotten, I bent over and slid my fins onto my feet.

The time had come to don the behemoth that is the BCD, tank, and regulator assembly. I started wriggling one arm into my vest, struggling until the boat captain came over and loosened the shoulder strap. He shot me a toothy smile and refrained to say what I'm sure he was thinking—that I had just pulled a serious rookie move. Right, you loosen the straps

as part of the gear assembly process, long before you sit down
to put the whole mess on. Once I was in the vest, I leaned
forward, wrapped the belt referred to as a "cummerbund"
around my waist just above my weight belt and stuck its two
Velcro parts together. Then I buckled the plastic closure over
it, snapped another together at my sternum, and tightened
the shoulder straps that the captain had just loosened for me.
I inhaled and pulled my mask up, pressing the eye cup to my
face first before arranging the back strap around my ponytail to
keep my hair from getting tangled. Finally, I exhaled and gave
the captain, who had watched all of this, a nod.

"Lista?" he said. Ready?

As ready as ever, I thought. But what I said was, "Claro,"
trying to sound like I did this all the time.

"I've got your tank. Stand slowly." I had forgotten how
heavy a full air cylinder was: thirty-five pounds, and an awk-
wardly shaped thirty-five pounds at that, especially when added
to the ten pounds hanging off of my waist. It was a good thing
he was helping me up. "Now, turn around and shuffle to the
back of the boat." That, too, was awkward. Fins are hard to
walk in, though shuffling backward is much easier than lifting
the front of the fin up into the air like an overzealous duck
trying to avoid tripping over its own feet. Still, it's not easy.

In fact, nothing about this process was easy. With all of this
equipment, I felt like an alien in my own body. Did I really
want to be doing this? Wasn't all of this stuff—and the rigama-
role involved with assembling it—part of the reason I stopped
diving years ago? I just had to do it once, I told myself. If I
wasn't into it, I could spend the next several days swimming
and reading novels.

The captain offered me his arm and guided me to the back
of the boat, hanging on to my tank to make sure I didn't lose
my balance as the boat pitched in the swell. "Bueno. Sit down

on the edge. Inflate your BCD. Now, regulator in your mouth. Hand on your mask." I did as he said and took a deep breath from the air in the tank. It was flowing freely. "Over you go!"

My reverse cannonball must have been effective since, after splashing down, I found myself floating head up above the surface, thanks to the air I had put in my vest. I was in the water, at last, where the bulky mass of equipment on my back became weightless. All I had to do here was bob up and down, breathing through my regulator. The sea was a gorgeous combination of turquoise and deep blue, and the island's emerald mountains rose up dramatically behind the tiny sand beaches and mangrove colonies rimming the shoreline. I was perched atop an environment I knew and loved and was about to be completely immersed in it. I *did* want to be doing this.

I grabbed the mooring line to avoid drifting away from the boat. Then I waited and watched Daniel and the other divers back roll in, one by one. Right about the time I was starting to get a little seasick, Daniel flashed the "Okay?" sign. After everyone had flashed it back to him, he gave us a thumbs down—the signal to begin slowly descending. I grabbed my BCD hose, held it over my head, and pressed the button that empties air from the vest. Right away, I started to sink, feet first. In a matter of seconds, my whole head was underwater, and I felt the pressure change at the center of my skull, the squeezing sensation that accompanies airplane takeoffs and landings. I pinched my nose and blew to clear it, then sank a little more. Other than occasionally repeating this action, I just had to let gravity work for another minute. My own body weight plus the weight of my equipment were enough to drop me to the ocean floor thirty feet below.

I hit the sand and came down onto my knees to wait for the others. In that moment, any remaining nervousness I felt was immediately replaced with wonder. First, because I could breathe underwater. All I needed to do was keep the regulator in my mouth and inhale. How could I have forgotten how cool that is? Then, because everywhere my mask-covered eyes landed, they spotted a new movement, shade, or shape. To my left was a school of about twenty-five blue tang—neon-indigo, three-inch-long fish that skitter over the tops of the coral, flashing their little yellow tails like roadside flaggers as they swim by. To my right was a Nassau grouper, a foot-long fish graced with a glamorous brown-and-beige stripe pattern. But the creature that most powerfully pulled my attention was the giant barrel sponge sitting right in front of me.

"Giant" isn't just a descriptor for this animal who can grow to be six feet tall and six feet in diameter; it's also part of the common name for the invertebrate *Xestospongia muta*. I was drawn to it because it was, yes, giant, but also because it looked nothing like the fish swimming around it or the shrimp clustered beneath it. In fact, it didn't even look like a living thing.

The "barrel" part of this creature's name comes from the fact that many of them have narrow bottoms and tops separated by bulging midsections, much like wheel-thrown clay bowls in the moments before the potter opens them up. This particular barrel shape must have gone a little wonky in the potter's hands, however, since nothing about its curves was regular. And its surface texture reminded me of those wax-encrusted wine bottles that adorn the tables of old-school Italian restaurants. I instantly wanted to touch it to see if it felt like wax or flesh or the kind of sponge I have in my kitchen sink. But touching reef dwellers is a major no-no. I didn't want to break off a chunk of a creature who grows very slowly. And I really didn't want to break off a chunk of a creature who might be hundreds or even thousands of years older than me.

I knew what it would feel like anyway: hard, like coral. Before I started diving, I, like many people, had assumed all sponges felt, well, spongy, like the loofas my mother had kept in the shower when I was young. I soon learned that there are over eighty-five hundred species of sponges on the planet, and only a few of them are soft. Those few have been so overharvested that it's actually somewhat difficult to find them in stores today. Meanwhile, the rest of the world's sponges have hard exoskeletons made from a variety of mineral compounds. Giant barrel sponges are among them, so there's not much temptation to squeeze them.

I floated over this stately reef dweller, pausing to admire the tiny fish—a rainbow wrasse—hanging out inside the sponge's protective cavern. This hollow central space is called an osculum, and staring into it is like looking into a deep hole or an unlit cave: more like the interior of a landform than the interior of an animal. I wondered what it was like for the wrasse to be in there, in the dark, surrounded by sponge. I bet time slowed for the little guy, since it was out of the way of predators, off-duty from its usually busy schedule of parasite-cleaning tasks, and out of the ever-present current with its nonstop barrage of smells, tastes, and sensations. It occurred to me that the barrel sponge was providing this flashy little fish with the opportunity to take a rest, to step out of the flow of time.

As I watched the wrasse float motionless in the sponge's sanctuary, I rested too. In this pause, I watched and appreciated. And I considered the possibility that my life—and perhaps the lives of many busy Westerners—might benefit from doing this more often. This, too, I had forgotten. It had been such a long time since I put myself into this otherworldly reality.

I first started scuba diving in high school. My family had a tradition of going to a Caribbean island every February when my brother and I had a week off from school. I think these trips were mostly driven by my mother, a sun worshipper with an olive complexion who liked to escape from the New Jersey winters and recharge her tan. In the tropics, my father's pale Irish skin prevents him from going outside when the sun is strongest. My brother and I have inherited this trait, so there were three of us needing an activity that kept us in the shadows during the heat of the day. Enter scuba diving.

After a handful of years of enthusiastic snorkeling, the three of us signed up for a course that would earn us the classic entry-level scuba certification: Open Water Diver. Within a week, we had little laminated cards that allowed us to purchase air, rent equipment, and tag along with guides who led us around reefs, above wrecks, and along deep walls.

I instantly loved diving. As a swimmer, I was already at home in the water, and years of floating around in the Caribbean had introduced me to the diversity of life in tropical seas. Diving took that diversity to a whole new level, however—both because I was deeper in the water and because I was in it for longer. In time, I'd logged respectable lists of species I had spotted: queen angelfish, parrotfish, trunkfish, needlefish, groupers, hamlets, puffers, porcupinefish, gobies, barracuda, triggerfish, eels, turtles, lobsters, and even octopuses. I went back to our hotel rooms and buried my nose in books dedicated to reef-life identification, sharing fun factoids about blennies and cleaner shrimp with my family over dinner.

I also loved the diving way of being. Life in the water is different from life on land, in more ways than I had initially thought. For starters, I had to learn how to adapt to the awkwardness of moving through a substance thicker and heavier than air—and how to do so while wearing cumbersome gear. Like most divers, I did this by slowing down. Sudden movements underwater just don't work; they're inefficient and graceless. So I shifted to kicking my fins completely, forming hand signals deliberately, and turning my body slowly and mindfully.

Unlike in the terrestrial realm, where the air we breathe is one of the only free and infinite resources we have, diving requires that you carry all of your air on your back in a pressurized metal cylinder. Running out of this precious resource is either dangerous or deadly, depending how deep you are when it happens. Many safety measures have been put in place to make sure it never does—including the standard practice of ending the dive when one person in the group is down to about 800 psi in a tank that started at 3,000 psi. As a result, carefully allocating your air is a key component of this activity. You don't want to be the reason everyone has to surface early.

The best way to use less oxygen is to slow down your movements—yet another reason to scale back your pace. I routinely watched the divemasters who took us out: In addition to kicking their legs every three to four seconds at most, they clasped their hands to make sure they weren't tempted to use them for propulsion or useless gesticulation. They changed directions slowly, nodded slowly, and consulted their gauges slowly. I knew their breathing must be downright glacial.

I lived in New Jersey at the time, and every day I sped through a forty-five-minute commute in New York City traffic to attend an ultracompetitive high school. There, I ate my lunch in five minutes flat so I could do some extra studying in the library. I raced to my classes to get a good seat and ran from the dinner table to get back to my homework. Then I went to bed and set an alarm for the latest possible moment that would still allow me to get up and get to school on time. Being forced to adapt to the liquid world was a blessing. It was the only time I ever consciously slowed myself down, the only time I got to step out of time. In that sense, it felt like the only window I had into another way of being—one that felt almost antihuman.

As I slowly and deliberately moved my fins to stay within sight of Daniel, I saw sponges everywhere I looked. The giant sponges were the most obvious of them; their size and shape stood out against any backdrop. But there were myriad other varieties as well: brown tube sponges and pitted tube sponges, and some purplish-red cylindrical clusters called stovepipe sponges. There were pink vase sponges, who look like their name suggests, and rope sponges, who stick out from the reef like errant pipe cleaners. A wild array of encrusting sponges smothered rocks and chunks of coral like lichens colonizing a downed tree. When I thought back to my dives from decades earlier, I had no memory of sponges being such dominant members of the reef ecosystem, but in all likelihood I was too busy chasing faster, flashier species back then to notice. The incredible array of sponge colors, shapes, and textures sur-rounding me that day in Roatan made it clear that they were important citizens of the coral reef community.

Sponges are filter feeders, which basically means that they move water through their bodies (in through one set of pores, out through another, propelled by tiny flagella) and eat whatever minuscule food particles—plankton, predomi-nantly—float their way. In addition to providing nourishment for the sponge, this process maintains water clarity, eliminates harmful bacteria from the surrounding sea, and keeps oceanic algal infestations down. Because they can filter up to twenty thousand times their body weight in water every day, they're basically the ocean's wastewater treatment plant.

There's a popular YouTube video that shows this in action. In it, a diver squirts a nontoxic neon-yellow dye around the base of a giant barrel sponge and then moves away to watch what happens. The color allows the viewer to see the water being drawn into the sponge's body and then extruded out its osculum. I cringed as I watched the diver touch the sponge, but once the dye started steaming out the top of it, I was riveted. "It pours out like smoke from a chimney," the narrator says as the diver continues to shoot dye out of a syringe. He's right; that's exactly what it looks like.

In the process of filtering the surrounding water, sponges also isolate and excrete elements such as carbon, nitrogen, and phosphorus that other reef dwellers need for survival. Corals release these organic compounds into the ocean, and sponges basically "recycle" them when they shed their filtering cells. These cells fall to the seafloor where they're eaten by snails and other invertebrates, who in turn are eaten by fish and larger predators. Some scientists consider this process so important that they refer to it as "the sponge loop" and claim that a functioning sponge loop is a critical component of a healthy reef ecosystem.

There are a couple of animals who earn the fabulous title "spongivore" (sponge eater), including hawksbill turtles, rock beauties, and spotted trunkfish—all creatures I was also seeing through my mask. But other than these few species, sponges don't have many predators and can live to be quite old. In fact, many people consider them to be the oldest animals on Earth.

I didn't get to see the biggest and oldest giant barrel sponge in Roatan, unfortunately. It occupies a site that locals call "Texas" (as in, "everything's bigger"), which is a deeper and more remote locale than someone returning to the sport after twenty-five years should be exploring. However, divers are notorious for taking underwater selfies with the Texas Sponge, so I've seen a handful of these impressive portraits online. From the looks of it, a medium-sized child would easily fit inside this giant. It also has an especially varied texture that I want to call a topography, as there appear to be ridges and valleys covering its surface. The Texas Sponge is remarkably symmetrical, compared with neighboring sponges that grow into a tilted posture, leaning with the prevailing current. But it's the osculum of the Texas Sponge that really stands out to me. It thins at its edges and flares outward, opening in something like a welcoming gesture, despite its incredible foreignness.

If the Texas Sponge is indeed one thousand years old—a good rough estimate—it's a relative youngster when compared with a well-known giant barrel sponge that lived near the Caribbean island of Curaçao. Scientists believe that this creature, who died in 2000, was about twenty-three hundred years old. When Hannibal was crossing the Alps en route to Rome and when the Great Wall of China was being built, this sponge was beginning to establish itself, beginning to contribute filtered water to its corner of the ocean. It kept growing through the fall of Rome, the rise of Christianity, the expansion of Renaissance culture, and the exploration and exploitation of the Western Hemisphere by Europeans. It consumed protozoa all through the decimation of the Aztec, Maya, and Inca peoples, the invention of the steam engine, and the creation of the automobile, airplane, and worldwide web.

The twenty-three-hundred-year-old Curaçao barrel sponge did not die of old age; it contracted orange band disease. In the span of ten days, it changed color, stopped moving water through its body, and wasted away. No one has been able to identify the source of this disease. As a result, scientists suspect that increased ocean temperatures or marine contaminants may be the culprits. We've raised the ocean temperature in less than fifty years. That number represents a tiny blip in the life of a sponge, but it's roughly the length of my lifespan so far.

Even if I were able to spend the rest of that lifespan with a tank strapped to my back moving at one-tenth of my normal speed, I don't think I would understand what time feels like to a sponge—if it even senses time at all. We humans are so aware of the passing of our hours, years, and lives that we cannot imagine a life outside of time. But my guess is that a sponge lives one.

As I hovered over examples of an animal family whose members include creatures who can look like everything from orange ooze to yellow volcanoes and from white goblets to brown mushroom spores, I felt like their timelessness might be just a little bit contagious. It seemed like doing an antihuman activity in an antihuman environment in order to be in their presence was giving me a glimpse into antihumanhood.

Because underwater reality is so foreign and hazardous, we have invented dive computers to tether us to the human world while we're away from it. Most divers wear one strapped to their wrist. These oversized watch-like contraptions serve a variety of purposes, but the most important one is to keep track of how long you spend at various depths. The computers plug that data into algorithms that estimate the amount of nitrogen accumulating in your blood—a natural and inevitable part of consuming compressed air at pressures greater than what we experience at sea level. Some nitrogen is not a problem, especially if you ascend slowly and spend a chunk of time on the surface between dives. Both of these activities allow much of the gas to dissipate into the air or water. Things get dicey only when you allow too much nitrogen to get into your blood and stay there. Dive computers were designed to make sure you never do this.

Time is an integral part of these calculations, so not sur-prisingly, all dive computers keep track of and display time in a variety of configurations: clock time; time underwater; time you can remain at your current depth without putting yourself in danger; and when you're on the surface, time elapsed since your last dive ended.

I didn't have a dive computer in Roatan. Because I was always with divemasters, any guidelines they followed would more than cover me if I stayed at more conservative, shallower depths. Without a computer, I had no way to measure time underwater. This turned out to be a blessing. I have no idea how long I lingered looking in the eyes of the green moray eel at Jager Bomb. I'm not sure if I met the eyes of the huge leatherback turtle for seconds or minutes. I think I let a school of blue chromis surround me for only a couple of breaths, but I really have no idea. I was on dive time, moving like molasses through an ecosystem that my species did not evolve to inhabit. And I was loving it.

On every dive, there's a point at which the person with the most rapid oxygen consumption signals the divemaster to indicate they're getting low on air. This usually prompts the divemaster to let the rest of the group know that it's time to ascend to fifteen feet, the depth at which it's customary to take a "safety stop." A safety stop is a three-minute pause that allows nitrogen to leave the body before surfacing, just in case you actually went deeper or stayed longer than your instruments indicated was safe. This is the moment when time kicks back in for me.

At the safety stop, everyone starts staring at their computers, as if watching the numbers count down will somehow make those three minutes go faster. As if we want them to go faster. As if we want to leave this amazing world. It's a sad moment for me. The reappearance of human time reminds me that I can't stay in the ocean, that I don't belong there. I am not a sponge. In fact, I am an antisponge.

Most humans consider our species to inhabit the top of the evolutionary ladder. Sponges, on the other hand, are usually relegated to the bottom rung. They are thought to be one of the most primitive animals on Earth. They don't have respiratory systems or digestive systems or circulatory systems. They don't have organs, they don't have tissues, and they don't even have specialized cells—which is to say that every cell in the body of a sponge can do every function a cell might need to be able to perform to keep the organism alive. We know this because sponges that have been liquefied in laboratory blenders have re-formed afterward. Their cells found each other and did what was necessary to make the creature functional again. They are the antithesis of human cells, which start diversifying and differentiating three weeks after the fertilization of an embryo. There are only specialists in our bodies, no generalists.

Generalization must be an effective evolutionary strategy, though, since barrel sponges have populated our planet's seas for at least five hundred million years—a number I can hardly conceive of. What helps me better wrap my head around the duration of the barrel sponge's existence on Earth is to think about what else was going on five hundred million years ago: the Cambrian explosion. This is the wedge of the geological time scale during which life on Earth transitioned from simple single-cell and multicellular organisms to a vast array of complex plants and animals like the ones we still see today. Sponges were one of the phyla that emerged from this burst of diversity, and they have not changed all that much since then.

In contrast, anatomically modern human beings have existed for three hundred thousand years. That's about one-one thousandth of the time barrel sponges have been around. We've changed a lot during that short period, however, and we've changed our surroundings even more. We have driven to extinction everything from saber-toothed cats, woolly mammoths, and mastodons to, more recently, passenger pigeons, golden toads, and black rhinoceroses. We have altered the shape of the seafloor. We have disassembled mountains. We have pulled water up from the depths. We have drilled miles into the planet's core, extracting the crude oil produced by the decomposition of million-year-old marine creatures. In burning their remains, we have altered the composition of the atmosphere and changed the temperature and acidity of the oceans.

Imagine if barrel sponges had been doing this for five hundred million years. They haven't, of course. Instead, they have been giving back to their surroundings, cleaning the water, and recycling nutrients day after day for millennia.

Sadly, this is another way in which they are antihumans.

There are two kinds of safety stops. When you're diving from a boat rather than from shore, the open-water safety stop is more common. In this case, you're at, say, fifty feet when it's time to come up for your three-minute layover at fifteen feet. Unless you're next to a reef wall, there are no markers around you, and the bottom—along with most of the marine life—is out of sight down below you. You either look at your depth gauge and use it to help yourself stay in one place or you grab a mooring line or an anchor line if one is available. Regardless, you're hovering in the deep blue, often with nothing much to look at other than your dive computer and the masked faces of your companions.

The other kind of safety stop is the shallow-water one. If your dive happens to end in water that is about fifteen feet deep, you don't have to float in the void; instead, you get to hang out right above the reef and watch the show for another three minutes. The divemasters I spent time with in Roatan, in general, liked to swim around a lot during their tours. This is understandable; you see more of the reef that way. But it means you're always moving, even if that movement is slow. As a result, when I had the chance to just stop and look around, I cherished it.

That day at Jager Bomb, my first day back to diving, we gradually meandered into shallower water as time ran down, following a trail of staghorn coral back toward the mooring line. Then Daniel flashed the safety stop sign, the palm of his left hand touching the tips of his first three right fingers. I adjusted my buoyancy to make sure I could hover in place for three minutes without touching any of the fragile organisms around me, then I crossed my arms in front of my chest and looked around. I didn't need to move because there were so many things moving around me: pale red squirrelfish with their giant black eyes, schools of smallmouth grunts with their horizontal stripes, and a couple of queen triggerfish with their permanently puckered lips and 1980s-style DayGlo scales.

I froze altogether when about thirty sergeant majors swam by. They're a common species on Caribbean reefs, but they never cease to draw my attention. They have lemon-yellow patches on their backs that fade to a pale gray along their sides, and those sides are marked by five solid black vertical stripes, like their military namesakes. When sergeant majors swim in large groups, the combination of the flashing caution lights on their tops and oscillating iron bars on their bodies make for a fun visual spectacle. If they're in the area, the best thing to do is stop and take them in, which is exactly what I did.

As I watched the sergeant majors glide over the reef and out of sight, I turned my eyes back down to the seafloor, which, of course, was covered with sponges of all types. One particular barrel sponge caught my eye. While many barrel sponges have that characteristic dripping wax texture, their surfaces can have other features as well. The sponge below me at the safety stop had ridges that looked almost like fins—or better yet, gills. The knobby protuberances lined up vertically, creating deep fissures that ran from the sponge's base up to the edge of its osculum. It actually resembled the cabin air filter I had recently replaced in my car, which of course served to remind me that, although there was no dye to help me see the process in action, the sponge was performing its community service right before my eyes.

I hung out in my spot for three minutes, trying to stay as still as the sponge, trying to fully appreciate everything about it—from its brownish-red color and unique mounds and crevices to its thin-lipped opening and lopsided lean. That level of focus made time stop.

Was I an antihuman for that brief moment? And if so, would I be able to bring that perspective with me as I ascended to the human-dominated, oxygenated world?

When Daniel's computer indicated that three minutes had passed, he gave me the thumbs up sign—the signal to ascend slowly to the surface where the boat was waiting for us. I gave him the "okay" signal in return, but I didn't head up right away. I looked back down at the barrel sponge and then kicked my way over to it so that I was floating just above its osculum. Then I swished my fins gently and imagined that I was rising on a column of water, one that emanated from the center of the sponge. I rode it slowly up to the surface where I removed my regulator from my mouth and took a deep breath. I pursed my lips and blew air through them, making my exhalation last as long as possible.

NEGOTIATING FLUIDITY

We are dodging icebergs at twenty-five miles per hour. From the bow of our twenty-eight-foot Zodiac, I try to make sense of the ecosystem I've come here to investigate: northern Alaska's Beaufort Sea coastline. But my customary visual bearings don't seem to be serving me here in Alice's Arctic Wonderland, where even the most fundamental rules of spatial arrangement have been upended. I see liquid lying over land, tundra hovering in midair, and chunks of ice floating several feet above the sea. I strain to delineate boundaries between water and sky, solid and gas, near and far. Where I expect borders, I find continuity— gradations of color, shifting shapes, and fluid forms. Reflections are sharper than the objects that make them, forcing me to question which way is up.

As we slow down to make a turn, I study the nearest iceberg like I would a puzzle piece, attempting to snap it into an orderly universe—the universe I am used to, the one I thought I lived in. But just as I do so, its contours morph. Its gently sloping side becomes serrated. From its flat profile emerges a third dimension. Shaking my head and intentionally blurring my vision do nothing to organize the landscape. I decide that I'm having an ice-induced hallucination.

In an effort to anchor my perceptions, I focus on the steady, comforting hum of a 115-horsepower Evinrude at full throttle. It lulls me into a half-sleep, an eerie liminal state in which I melt into my surroundings. My surroundings are, themselves, melting because we are in the Arctic. Northern Alaska is one of the parts of our planet that has begun to experience the tangible effects of human-induced climate change. Of course, the primary cause of human-induced climate change is our species' use of fossil-fuel-powered internal combustion engines, like the one I'm sitting next to. In addition to delivering me to this trippy northern netherworld, it emits carbon dioxide, the gas that has changed everything. It has affected the amount of ice that surrounds our boat and the nesting behaviors of the birds we are here to study. It has affected weather patterns, storm cycles, migratory routes, and the speed at which extinctions are occurring on this planet.

I had been looking forward my time in the Arctic National Wildlife Refuge since a cheery government employee called to tell me I had been granted a Voices of the Wilderness artist's residency. She asked me if I wanted to accompany a group of biologists to this rugged and remote parcel of land I first heard about as a teenager, when I spent the summer of 1989 hiking and kayaking in Alaska. With no hesitation whatsoever, I bought a ticket to Fairbanks and blocked out a two-week chunk of August.

An hour into our six-day research trip, however, I had already begun to feel both a mind-bending disorientation and a disturbing complicity.

"Here? They build nests here?" I exclaim. After some video-game-like high-speed iceberg avoidance, we emerge into a zone of open water and land at our first gravel bar. I look in astonishment at a flat, featureless, and lonely landform composed of nothing more than an accumulation of half-dollar-sized rocks. This is the kind of terrain where female common eiders—the

migratory waterfowl we're here to study—build their nests.

"She's sitting on her eggs—right up there, by that group of glaucous gulls." Will stands at the console of our boat, peering through his binoculars.

"Okay. Noose or net?" Elyssa asks from her position in the back of the other Zodiac.

"Noose, I think. This island is too narrow for the net. If I approach from the lagoon side, I think I can get her."

"Cool. Bridget, can you pass the noose pole over to Will? And grab the banding kit."

Elyssa Watford, Will Weise, and their US Fish and Wildlife Service crew spend their summers looking for common eider nests. Their mission is to gather data about female eiders' nest-site selection and the birds' physiological responses to their nesting choices. Although I'm technically a "visiting artist," I'm really just an extra set of hands and eyes. Only fifteen hundred or so people experience this 19.5-million-acre preserve every year. I'm thrilled to be one of them—especially one that has the potential to do some good, to further our understanding of this species and the environment it occupies—so I'm more than willing to do whatever grunt work is assigned to me.

Without an ice barrier for protection, a big tide—or even a moderate storm—could wipe out this sand spit, along with the shallow depressions in which these birds lay their three or four sage-green eggs. But these precarious sites have advantages that, for now, outweigh the costs of migrating all the way up here to northeast Alaska from the eiders' more hospitable home waters: the Bering and Chuchki Straits of southwestern Alaska, where they spend all but three months of their year. The main benefit of this site is access to the all-you-can-eat invertebrate buffet that the Beaufort Sea becomes during the long days of the Arctic summer. Newborn eiders need only walk a few feet toward the water to get their first meal of mussels or clams.

This advantage might not last, however, as weather patterns shift in the Beaufort Sea.

"Typically," Will says, "we've got the polar ice cap breaking up and creating large flooding events and nearshore storm surges in August or September." That's after the babies have fledged. "But these days," he continues, "we're starting to see more of that earlier in the season, like in July." Earlier and more powerful storm surges make laying eggs by the waterline riskier. To avoid the threat of swamped nests, it seems likely that mother eiders will start choosing to bed down in higher, drier areas. Because these areas are more exposed to the frigid Arctic winds, they'll require the birds to burn more calories to stay warm. Elyssa, a master's student at the University of Alaska, is working to understand how these different nest-site choices might affect the birds' metabolic processes. In other words, she wants to know just how much harder incubation is going to become as the effects of climate change increase.

Because of its low-lying topography and proximity to the North Pole, this ecosystem is showing evidence of the effects of climate change earlier and more dramatically than others. According to a US Fish and Wildlife Service handout I picked up back at the Fairbanks office, "The Arctic is warming at a rate almost twice the global average." This is largely due to ice-albedo feedback—a fancy name for what happens when light-colored ice melts away and exposes darker earth. The white stuff reflects heat, but the dark stuff absorbs it, exacerbating the warming and melting effects of the sun once ice no longer covers the ground. The same little booklet went on to say that the area in which we're traveling—Alaska's North Slope—is expected to be 12.5 degrees Fahrenheit hotter by the year 2100. In order to survive, the creatures that dwell in this constantly changing landscape will have to adapt to radically different conditions.

At some point, breeding in the Arctic might not work for common eiders. "That's the biggest reason these birds are considered vulnerable," Will tells me, as he unscrews the lid to a red plastic gas canister. I flinch when the smell of benzene hits my nostrils. We've already burned through a tank of gas. "If those islands change and disappear," he continues, "well, that's where they nest. Who knows if they'll find somewhere else to lay their eggs or not."

To predict what the eiders might do, Elyssa needs data. To get data, we've got to brave the iceberg-strewn Beaufort Sea. Our six-person crew left Kaktovik, an Inupiat village of 250 people, in two motorized inflatable boats stocked with eight large containers of fuel. We have multiple dry bags full of minimally invasive data collection tools, lightweight tents, propane stoves, canned and dried food, and buckets in which to deposit and carry out our human waste—equipment aimed at keeping us self-sufficient and reducing our impact on the land.

A lot of resources have been invested into preserving the notion that our actions don't affect this ecosystem.

After ten minutes of motoring, I've been transported from the no-nonsense, nuts-and-bolts process of data collection—the world of boundaries, boxes, measurements, and thresholds—to the borderless, oscillating iceberg dreamscape. I peer out from behind my polarized sunglasses, squinting at the fuzzy conjunction of water and land. At 70 degrees north latitude, where the world hovers between liquid and solid states, the horizon is not a line but a zone. Quantum physicists might call it a "field of probability." The land and sea collide with the sky somewhere in the range of my sight, but unearthly mirages make it impossible to determine exactly where and when that meeting occurs.

I've seen mirages before, in the desert Southwest, when the asphalt heats up and ripples my windshield view. The illusion lasts for a minute or two, then my conventional reality reasserts itself, and the highway falls back into place. Arctic mirages are altogether different. Every visible landmark—hillock, ship, iceberg, building—appears to float untethered in the sky. No matter how fast we go, no matter how close we get to these objects in our sight, they resist settling back into their expected arrangement. I gaze at the weird and warbly horizon until I can no longer tolerate its oddity.

"Fata Morgana" is the lyrical term used to describe mirages that occur at extreme latitudes like northern Alaska's—mirages that distort objects so drastically they become unrecognizable. The phrase has a medieval ring to it because of the reference to Morgan La Fay, infamous sorceress, student of Merlin, and half sister to King Arthur. She was said to lure sailors to their deaths with visions of ghost ships and phantom landmasses.

This optical illusion requires the presence of a thermal inversion, which is to say, a condition where the temperature of the air close to Earth is colder than the temperature of the air above it—the opposite of what is normally the case. It also requires the existence of an "atmospheric duct," a horizontal layer that essentially traps light waves, causing them to follow Earth's curvature rather than travel linearly. Some waves are reflected, some are refracted, and some are focused. The process distorts, displaces, and inverts images in ways similar to a prism or a cut-glass pendant in a sunny window.

On the Beaufort Sea, conditions are ideal for the production of Fata Morgana. The preponderance of ice keeps the temperature of Earth's surface uniformly low while the brightness of the sun warms the air above it. The temperature affects the light, and the light affects my vision.

My vision is the sense I rely on to make sense of these peculiar surroundings. It's unnerving to think that it might not be trustworthy.

While Elyssa assembles the equipment needed to measure, weigh, and draw blood from the mother eider, Will walks slowly in thigh-deep, 38-degree-Fahrenheit water. He's shed his Mustang suit, the foam-filled, full-body, electric-orange safety garment that each of us wears on the boats at all times. Theoretically, these mobility-reducing getups will keep us afloat if we happen to fall overboard. Under them, we're wearing waterproof fishing waders and, under those, several wool shirts and fleece jackets. This layering system is required to maintain a reasonable body temperature while working in a windy, rainy environment where the ambient air temperature hovers between 35 and 45 degrees Fahrenheit.

As Will approaches the nest where the female sits incubating her eggs, he extends the ten-meter telescoping rod they refer to as a "noose pole," a term that I eventually embrace, despite its morbid overtones. It's a device originally designed for carp fishing that the team has jury-rigged for the specific task of catching nesting eiders. Female eiders look nothing like carp; in fact, their coloring is rather dull. While male eiders sport striking black-and-white patches accentuated by electric-orange beaks, these ladies look a lot like their distant cousins: female mallard ducks. Both share plumage that celebrates the world's infinite variety of tans, beiges, and browns, providing them with camouflage. Eiders are much bigger than mallards, however. The average female weighs about four or five pounds, a size Elyssa describes as "halfway between a duck and a goose." I peer at the mother eider through my binoculars, thinking how much I would like to hold her. She strikes me as beautiful and powerful, and I think that it must be magical to experience her presence through a physical connection.

I watch as Will dangles the end of the pole over her head and slowly lowers the circle of monofilament toward her beak.

When he slides the tool toward himself, the noose tightens around the duck's neck. I shudder, even though I know he's using only the amount of force necessary to keep her from getting away.

If Will had startled and flushed the female, we would have moved in toward the temporarily motherless nest to collect data on both its placement and the status of the eggs it contains. We would pick up each one and look through its shell to estimate the age of the embryo inside. We would mount a camera nearby to capture a few days of footage of the mother's incubation and stick a fake egg—one that encloses a tiny microphone wired to a recording device—into her nest to measure her heart rate.

But since Will's got a live mother in his hands, we have the chance to collect more intimate information about her—data obtained through mouth swabs, fecal samples, and blood draws. The results will yield information about the bird's metabolic processes, fueling Elyssa's master's thesis on the eiders' physiological responses to their nesting choices. Her conclusions will contribute to our pool of baseline data, allowing scientists to quantify the degree to which human actions are affecting the eiders' lifestyles. Of course, the idea of "baseline data" in an ever-shifting reality seems as useful as my attempts to fix icebergs into an orderly visual arrangement. And I can't help noticing that every bird on this sand spit—not just the eiders—has taken to flight for the duration of the time we spend onshore. The cacophony of their disgruntled squawks is overwhelming.

Nearly everyone knows that sea ice in the Arctic is on the decline. Many of us have seen pictures of polar bears stranded on isolated floes, and some of us may have heard that sailing the infamous Northwest Passage has gotten substantially easier. The entire nine-hundred-mile route from the Beaufort Sea to Baffin Island was ice-free for the first time in 2007, and in 2016, the first cruise ship successfully navigated the entire journey.

Sea ice has always undergone some melting during the summers when the sun is most powerful. How much melting is directly related to how much sunlight it reflects. Because of the ice-albedo effect, about half of the solar radiation that falls on sea ice is radiated back into the atmosphere. The other half is absorbed. While this absorption rate is less than that of ocean water, it's enough to liquefy the ice. If sea ice is blanketed with snow—a highly reflective substance—it absorbs less solar radiation and takes longer to melt. If that snow melts, however, the puddles of water that remain absorb additional heat. The more heat the water absorbs, the more it melts the sea ice, producing more water—which in turn absorbs more heat, which produces more water. You can see where this is going. Scientists call this a "positive feedback loop"—a phenomenon in which changes to a system are amplified, increasing instability. Sea ice melting is a quintessential example of this, and it's accelerating dramatically in the Arctic right now.

I'd heard all of this many times; but until I looked at actual graphs, the severity of the situation had not quite hit me. Scientists from the National Snow and Ice Data Center (NSIDC) have been tracking Arctic sea ice coverage annually since 1978. When their data are plotted, they show a distinctive downward trend—at an angle that, ironically, resembles your average intermediate ski slope. In raw numbers, Arctic sea ice in September 1979 covered about eight million square kilometers. In 2012, the area had dropped to under four million square kilometers. That's like cutting the country of Brazil in half over a thirty-five-year period. Based on their models, the NSIDC scientists concluded that "Arctic sea ice extent will continue to decline, the eventual outcome being an essentially seasonally ice-free Arctic Ocean." An ice-free Arctic Ocean strikes me as an oxymoron. If the emblematic icebergs melt away, we will need a new way of thinking about that body of water. We might need to give it a new name.

Nomenclature aside, there is no question that the boundaries of Arctic Ocean travel will expand. Ships will enable us to fish areas that have not been fished before, drill for oil in places that were previously inaccessible, and deliver tourists to scenic locations that have, until this point, been visited only by well-meaning researchers in big orange flotation suits.

"Do you want to hold her?" Elyssa asks. I shake my head. I don't feel ready yet. Yes, I'm afraid of hurting her, but there's something else holding me back—something about having such a personal and direct impact on this wild creature. A logical voice inside derides my decision: "Everyone else on the crew is touching her. What's another set of hands?" But they are convinced that we belong here, that the good we're doing outweighs our intrusions into the birds' lives and homes. I'm less sure, and my uncertainty has me trying to locate this dubious ethical boundary.

Will grabs hold of the mother eider and nestles her into his lap in one smooth and efficient move. "Then pass me two bands," Elyssa says to me. She crimps them loosely, one around each leg.

"Calipers now?"

"Please." She grabs the tool, opens it, then closes it down—first around the bird's lower leg, then across the length of its beak, then along its head. "Tarsus 61.5. Culmen 51.9. Head 127.8."

"Scale coming."

"Good. 2,040 grams."

"Envelopes for the feathers?"

"Yeah." She pulls a few from the mother's head. "Nest ID and band number on the envelopes."

"Got it. Ready for the swabs?" I ask.

Each of these procedures has a distinct purpose. The banding enables future identification of these birds—not just by Elyssa and Will but by other scientists who might spot them. The measurements create a reserve of information about nesting females' "normal" sizes and weights, and the fecal and mouth swabs allow testing for disease. The feathers—all of which will molt later this summer, when the hens are busy raising their young—are retained for genetic and isotope analysis. The blood samples Elyssa extracts will be assessed for chemical compounds that yield information about the birds' nutritional status and levels of stress during different stages of incubation, the twenty-six-day chunk of time these mothers remain seated on their nests.

"The fact that these ladies fast and sit on their eggs 99.5 percent of the time—for twenty-six days—is just wild to me," says Elyssa. "It's amazing that they are able to starve for that long and still hatch their babies." This phenomenon is called "capital breeding." While eiders are not alone in employing this strategy—desert tortoises are well-known for it—the extreme duration of their fast is unique in the bird world. The longest time the mothers will spend off of their nests is the time it takes for us to poke and prod them.

Elyssa's lips purse as she sinks a needle into the eider's neck. I know she's successfully performed this procedure hundreds of times, and I understand her explanation for why it's necessary. Still, I feel uneasy, and I wonder if she does, too. The mother bird squirms in Will's arms, her webbed feet pedaling in the air, as though, with the right amount of effort, they might propel her away. After a few seconds of struggle, she settles down. I watch the syringe fill with her blood—dark red, like mine. She stares at me through tiny, distant eyes.

At the end of our second day of motoring, when we pull into Demarcation Bay to establish a base camp, we spot a double-masted ketch anchored out in this protected harbor near the Canadian border. I don't believe my eyes; not only am I seeing human beings in this remote corner of the world, but I'm seeing them aboard an old-school sailing vessel, one we immediately dub "the pirate ship." That night, the erstwhile pirates come to shore in their dingy to make a driftwood bonfire on our beach. We join them in their smoke-induced mosquito-free zone, and they invite us aboard their ship, the *Infinity*, for dinner the next day. In addition to seeing the visit as an opportunity to explore the inside of this mysterious non-petroleum-powered craft, we're excited to dry out, warm up, and eat someone else's food.

After answering questions about eiders and Zodiacs and bear camping protocols, we scarf down freshly baked bread and steaming lentil dahl with the *Infinity*'s twenty-two crew members—all of whom are volunteers and some of whom have no previous sailing experience. This "cosmopolitan cast of characters" (their term) is led by Swiss captain Clemens Oestreich, who lives aboard with his wife and two young children. As we shed a few of our layers, they outline their mission: to raise the "EarthFlag" on the ever-shrinking Arctic ice cap. Captain Clemens explains to us that the ocean-blue flag depicts the "seed of life," a sacred geometric symbol of unity and balance, and that it has been proposed to serve as a banner for our planet.

According to the EarthFlag Foundation, the group that created the design and granted its rights to the commons, the symbol is "a reminder that everything is interconnected."

One of the crew members darts out of the dining hall and comes back with the flag stretched wide in his arms. I'm immediately struck by it. Unlike national flags—symbols of independence, borders, and separation—this one feels inclusive: seven interlocking white circles coming together to make one elegant, symmetrical emblem. The crew plans to raise the flag on a totem pole carved by residents of the island of Vanuatu in the South Pacific—the area they had embarked from six months and over six thousand miles earlier. They hope that by filming and distributing a documentary about their adventure, they can highlight the need for interdependent, multinational behavior and decision-making to reverse our planet's environmental trajectory.

I sit silent on the wooden floor. *They're crazy*, I think to myself. Crazy, but inspiring. I wonder if anyone will buy their film, if anyone will watch it, and if anyone will shift their outlook as a result.

The whole time we are camped in Demarcation Bay, the *Infinity*'s crew sits offshore monitoring various ever-changing models of iceberg distribution in the Northwest Passage. They plan to wait until the sea ice has broken up enough to allow them to make eastward progress toward their goal, well aware that warming air and water temperatures in the Arctic—direct fallout from our climate crisis—make their mission possible.

Since we've set up all of our "incubating eider cams" and placed all of our heart rate monitors—the objects I have come to think of as "imposter eggs"—Elyssa decides that we should survey some outlying sand spits for active or recently abandoned nests. "We've got about thirty kilometers of motoring ahead of us," she says from behind the gas tank. "Go ahead and layer up for the ride." I yank my neck gaiter up and tuck it under the lower rims of my sunglasses. Then I pull all three of my hoods over my head and close my eyes.

After some amount of time—who knows how long, since time, like light, seems to bend and curve here—the steady rush of icy images gives way to blackness, and some part of me recognizes that I am asleep. Or maybe not asleep exactly, since I am still sitting upright with my multi-mittened hand grasping a lash strap, but certainly not awake either; I am beginning to see people and places that are nowhere near me in the physical world. A friend from home floats through my awareness, followed by a vision of my brother as a toddler, a scene from my high school cafeteria, and the line of palm trees I can see from my bedroom window at home. They are like holograms, existing in multiple dimensions. While I can't say I am "used to" this mind-bending sensory space, I am learning to relax into it and open myself to its insights.

When we arrive at the six-kilometer-long, one-hundred-meter-wide spit, it is enveloped in fog. We spread out and begin to walk, each person about thirty meters from the next, staring down at the ground in search of the telltale circular depressions that indicate eider nesting activity. When I look across the sand at my companions, I cannot tell them apart any more than I can distinguish one female eider from another; we all wear the same light brown waders, carry the same hefty backpacks, peer through the same black binoculars. I continue walking in a straight line, parallel to but distant from the others' trajectories, growing weary from the penetrating frigidity of the wind. When it dies down, I become aware of the distinctive sound of rushing water, making me wonder if I've been transported again—this time to one of my many trips down California's American River or a stretch of the Yampa in Colorado. But this is not a hallucination; as I look to my right, I see icebergs coursing by like rush-hour traffic. They are rolling landward with the tide, riding a current as constant as the Green River's flow through Utah's Desolation Canyon—only, I am not in Utah or California or Colorado. Or am I?

With every hour here, I doubt the reliability of my perspective more. Not only are my traditional binaries—up/down, liquid/solid, sky/sea, awake/asleep—failing me, but I'm uncertain about my measurements of time and space, about the lines that govern where I end and where another creature begins, about what we are accomplishing as a team and what we are accomplishing as a species, and about how we might get out of the mess we continue to make.

From farther up the spit, I hear an unusual guttural sound. It starts as a low rumbling and builds to an almost piercing rattle, shocking me back to the beach where I stand. I know that sound all too well, from my many years of living in eastern Idaho. "Sandhill cranes!" I scream, proud to finally have beaten my professional birder companions to the punch. I'm also relieved to know a familiar creature is somewhere nearby. We all gather by a pile of driftwood and watch the three elegant birds strut among the rocks, dancing in and out of the chilly fog. Their gait is predictable, their red eye patches and tapered beaks somehow comforting to me. We stretch our backs and share what's left of a one-pound bag of M&M's, by far the most colorful objects for miles. A chill comes over me when the cranes take off for some other remote beach. We're the only vertebrates around, again. For a moment, I had felt like I wasn't so out of place.

I stuff a granola bar wrapper into the pocket of my waders and return to solitary walking. For the rest of the day, we find nothing else but hoof prints in the sand—caribou tracks. "Why are they here?" I ask, wondering the same thing about us. We've been walking for over three hours and have yet to find a nest. "There's no food for them. No fresh water either."

"They're fleeing from the hordes of mosquitoes up on the tundra," Elyssa says. We've been doing the same thing; when the wind died down the previous night, I couldn't bring myself to leave the tent to pee, for fear of returning with ten to twenty bites from my one minute of flesh exposure. The permafrost— the layer of frozen ground that forms the base of the Arctic earth—is melting along with the sea ice. As the disintegration of the land produces more standing water, will the bugs get worse? It's hard to imagine that being possible. Will the caribou be able to reach these pest-free Shangri-La's when the water level rises? Will these sand spits even exist?

When I stretch out in my cozy down sleeping bag for the night—with the tent netting zipped tightly shut against the whine—I find myself wondering if I'm surrounded by eider feathers as I dodge icebergs in my dreams.

Keeping warm with eiderdown is an ancient tradition, thanks to its superior insulating quality. The tiny feathers have extra hooks in their structures that grant them higher insulating capacity than the more commonly used goose down. Birgitta Berglund and her team from the Norwegian University of Science and Technology study eider feathers found in Nordic Iron Age (570–1030 CE) graves in an attempt to determine their precise geographic origins and ages. Several Viking-era sagas mention the harvesting of eiderdown as well, so it's safe to assume that Europeans have been collecting this resource for over a thousand years. The Inupiat and other Alaskan indigenous groups have done so for at least that long.

These days, an eiderdown duvet goes for upward of six thousand dollars. Most of today's commercially harvested eider feathers come from Iceland, so those blankets are transported around the globe in fossil-fuel-burning airplanes and cargo ships. These vessels produce emissions that contribute to the warming trend that melts the sea ice and makes it harder for the eiders to occupy the breeding grounds where they produce their coveted down. If my sleeping bag is made from imported eiderdown, I've helped displace the birds we're here to study— in an effort to keep them from being displaced. My heart sinks. I know that every one of my purchases has a ripple effect that I don't often see. Thanks to this Arctic perceptual upending I'm living, those effects are now staring me in the face.

Despite the warmth and comfort of that sleeping bag, sleep, like land-sea borders, can be elusive in the Arctic. In the summer, the sun never really sets; instead, it crosses the sky like a long-distance hiker traversing the coastal tundra. At 2:00 a.m., when it's cold enough that the bugs have gone into hiding, I unzip the tent door to the sight of the big orb hovering above the water, casting a dusky-orange pallor over the peaks of the Brooks Range that lie in the distance, far beyond our tents and boats. We don't see this kind of glow often in the lower forty-eight, although, as the composition of the atmosphere continues to change, we might be witnessing more reddish tones on the horizon very soon. I think about the last sunset I saw at home in Santa Cruz, sitting on Cowell Beach after surfing until I could no longer see the incoming sets. Maybe it's the twenty-four hours of light that are throwing off my cycle, or perhaps it's the general blurring of boundaries that's eroded even the frontier between wakeful consciousness and whatever state it is that I plunge into every night. Regardless, I find myself lingering in that fuzzy liminal zone, thinking about what we're doing here.

Seven degrees Fahrenheit. This is the average global temperature change the planet will experience between now and the year 2100 if our species makes no significant lifestyle changes. That number will be doubled here in the Arctic. Our continued use of fossil fuels and our general consumptive habits will cause a warming trend that will almost certainly precipitate two feet of sea-level rise and the extinction of 40–70 percent of species on the planet. One of those species might be the common eider.

Our little team is trying to prevent that—by cruising around in gas-powered boats, wearing petroleum-derived polyester layers, eating prepackaged meals cooked over a stove fueled by disposable metal butane cartridges, and interfering with the eiders' daily lives. I close my eyes, unable to fathom the degree to which I'm woven into this mess, unable to get a grasp on how I might be able to extricate myself from culpability, unable to see the obvious or easy solution that I want so badly to exist.

In response, I burrow back into the safety of my one-person mummy bag and try to maintain the illusion that I'm doing good here, that the positive effects of my writing might offset the resource investment required to get me here. That the data we collect will be used to help these birds. That all of this running around the globe trying to understand eiders, melting ice, and storm cycles isn't just another attempt to impose human control over forces that are much more powerful than we will ever be.

For our last night in the field, we move camp to Anglan Point, a former DEW Line site. "DEW" stands for Distant Early Warning, the program established by the Department of Defense to detect incoming Soviet bombers during the Cold War. Once upon a time, there was a radar station here, but it was dismantled and hauled away on a barge, along with many tons of contaminated soil packed into giant orange plastic bags. The reclamation process left a wide and flat area of impact—a perfect campsite for a group of eider researchers headed back to their home base. I am relieved when Elyssa assigns me to set up camp while she and a few other crew members go out to retrieve the last of the nest cameras. These tasks are familiar rituals for me—tying trucker's hitches to anchor the tents, connecting the fuel tank to the stove, heating water to fill thermoses—and they ground me for a few moments, until I remember that I am camped at a Superfund site, and I look up to see the tundra floating above the water yet again, another Fata Morgana horizon.

There is another DEW Line site in Kaktovik. It's still being used, and a Defense Department employee remains on guard inside the futuristic-looking globe that sits in the center of town. Since it's the largest human-constructed object for miles around, we will use it tomorrow to guide ourselves back to the gravel bar that serves as the local marina. For me, the presence of these Cold War relics hovers somewhere between spooky and quaint. Many people once thought that our biggest threat was the massive country on the other side of this melting polar ice cap; indeed, some still do. Others have come to realize that the threat is the melting polar ice cap itself. And unlike the former Soviet Union, this threat is one that permeates borders and boundaries, entangling all of us in its pernicious web.

After this week, I've become more and more convinced the immediate and pervasive threat of climate change will need to be combatted with perceptual overhauls, not missiles and bombers, incremental policy initiatives, or even more science. Still, we maintain the illusion that we can hybrid-car, carbon-tax, and cap-and-trade our way to back to the lifestyle we've grown accustomed to. As I lift my mosquito headnet to sip my hot cocoa, I squint out at the water, straining futilely to make out a clear boundary between the sea and the sky. I wait for the rest of the team to return, listening for the reassuring sound of their motor.

A few months after I return from Alaska, I notice a photograph of the pirate ship in Elyssa's Facebook feed. It is a repost from the *Infinity*'s own page, one they had been filling with snippets of video in anticipation of what will be a Discovery TV series called *Expedition to the Edge*. Spectacular drone footage reveals the ship docked at the edge of the polar ice cap. The camera then zooms in on the team hoisting the EarthFlag on a raised mound of ice at global warming's Ground Zero. Apparently, they had smooth sailing through the Northwest Passage after leaving their mooring in Demarcation Bay. A smile spreads across my face.

I follow a chain of links to the *Infinity*'s website. There, a photograph of the bonfire we shared on the beach serves as one of their banner images. Another is a picture of a polar bear, with a quote by architect and systems theorist Buckminster Fuller superimposed over it: "We are not going to be able to operate our spaceship Earth successfully unless we see it as a whole spaceship and our fate as common." I sit back in my office chair, struck by how succinctly this quote sums up the perceptual shift Clemens and his crew want to inspire. I learn from the website that Clemens, having succeeded in the north, plans to next raise the EarthFlag as far south on the planet as possible and needs money to do so.

Is this how we move forward? Or is this just another example of our species' colonization of the planet? *Probably both*, says the voice shaped by my own Arctic-induced perceptual shift. I have to root for this expedition for its lofty ideals, cooperative values, and adventurous spirit—all of which I share. And yet, here we are again: human beings pushing into places I'm not sure we belong, affecting living and nonliving things in ways we both can and cannot measure.

When season 1 of their Discovery series is made available in September 2020, my heart sinks at its description: "A rag tag group of family and friends quickly learn the price of adventure when a voyage with friends unravels into a life-or-death crisis forcing all hands-on-deck." Human drama sells. The need for cooperation in the face of a global environmental crisis does not.

According to the *Infinity*'s website, the crew planned to leave for Antarctica in the summer of 2021. A dive into their social media feeds in September 2021 revealed that they were still docked in Amsterdam, fundraising.

Back in the boats on our final high-speed glacier dodging run to Kaktovic, my brain soon tires—first from attempting to force the elements into the orderly arrangement my brain prefers, then from the process of relinquishing control. I gradually let myself slip into the psychedelic reality of Beaufort Sea boating once more. As disorienting and disruptive as my petroleum-powered space-time travel may be, I have come to wonder if the slippery scene around me might be a preview of our immediate future in a rapidly warming world.

Quantum physics has revealed that nothing has ever really been separated from anything else, although our collective Fata Morgana—the perspective that sees bordered nations, individual species, and isolated ecosystems—tricks us into thinking otherwise. We—and by we, I mean human beings, nonhuman beings, nonsentient beings, and even the forces we cannot see or understand—are all inextricably entwined, we are told. Like it or not, our thoughts and behaviors are waves that interfere with the billions of other ripples we swim through.

As town comes into sight, my unreliable eyes have lifted the giant dome of the old DEW station up and off the surface of Earth. Thanks to Fata Morgana, even the largest and most expensive structure in town has been uprooted and subjected to renegotiation. While the motor behind me thrums, the level in the gas tank ahead of me plummets, and our emissions rise to join the canopy of greenhouse gas molecules that keep in our planet's heat, it occurs to me that this word—"renegotiation"—is somehow critical to my understanding of this place, this planet. We could renegotiate our relationships—with each other, with the land, with the ice, with the eiders. We could also renegotiate the way we see the world by seeing the whole instead of the parts. We could choose to act as though we were strands woven into an exquisite multidimensional artwork— one that has never before existed in this form and will never look exactly the same again. We could unite under a planetary flag that represents all global residents, human and nonhuman alike. Maybe this is the first step: acknowledging that our current everyday reality is an unsustainable illusion. When we can see through new eyes, perhaps we'll see new solutions. Or perhaps we're stuck with a species-centric vision, sharpened by hundreds of thousands of years of natural selection favoring human dominance. I don't know. What I do know, or hope, is that radical challenges to our established perspectives— ones that, like the Arctic landscape, force us to negotiate the discomfort and disorientation of fluid boundaries—might offer some kind of way through. Or over. Or around. Or beyond.

About half a mile from Kaktovik, details finally come into focus. The DEW Line's spherical building is actually anchored to the tundra. The boats are tied off to shore, at least until a storm surge washes away their moorings. There are no polar bears roaming the island where a whale carcass sits left over from last September's hunt; although next month, the iconic creatures will wander ashore in search of the food they can no longer find on the shrinking sea ice. Normality appears to reign as I help unload the boats before sprinting to the bunkhouse. I'm anticipating a glorious hot shower and a delicious meal of vegetables flown in on the last bush plane from Fairbanks. I will soon be clean, dry, well-fed, and on my way back to my day-to-day life in coastal California. There, I will ration water, watch for record-breaking wildfires, worry about the disappearance of my favorite beach, and witness the ongoing decline of the monarch butterfly population—all while continuing to drive my car to the market where I can buy organic apples grown in Chile. When I look at my life from the right angle, I see mirages everywhere.

As we pull up to shore, I realize that I never ended up holding a mother eider. Everyone else did; they cradled them and plucked their feathers and drew their blood and released them again and again. I looked into their eyes and held their eggs, but I didn't make direct contact. When I stop to tally up the myriad ways in which my actions affect these birds, it's clear that our lives are intertwined. My intentions were good; I just wanted to leave them alone, untouched by humanity. Or by my humanity, anyway.

As if that were possible.

The research described in this article and the author's experience were funded by the US Fish & Wildlife Service. Portions of this essay were first published as part of the post "Chasing Eiders: A Week on the Beaufort Sea with Arctic Refuge Bird Biologists" on the blog US Fish & Wildlife Alaska: Stories from Alaska by the US Fish & Wildlife Service.

RIPPLING LINES

"Whatever you do, don't put your foot down," said one Baja surfing guidebook. "The stingrays are notorious," claimed another.

But the descriptions that followed, with phrases like "a long-boarder's dream" and "rides that go on forever," made a visit to Punta Abreojos irresistible. The spot was touted as one of the most consistent breaks on the peninsula, and back in 2005, the area around it was entirely undeveloped, allowing for beach-front camping in sight of the waves.

I'd had a little bit of experience with stingrays, since they also live on Baja's bay side, in the Gulf of California, where the outdoor school I worked for ran sea kayaking and sailing trips. We occasionally saw their fleshy, diamond-shaped bodies emerging from camouflaged hideouts in the sand. They would shudder into action, then quickly scurry away, like foot-long alien hovercraft trailing thick, whiplike tails. Because those tails could deliver a venomous sting, I always taught my students to do "the stingray shuffle," an awkward but mandatory water entrance move. Foot shuffling not only stirs up sand, but it also produces vibrations that send the rays swimming rapidly in the other direction. They prefer to have nothing to do with us, so, given a little advance warning, they'll avoid an interaction. The stingray shuffle had successfully kept our worlds apart for years.

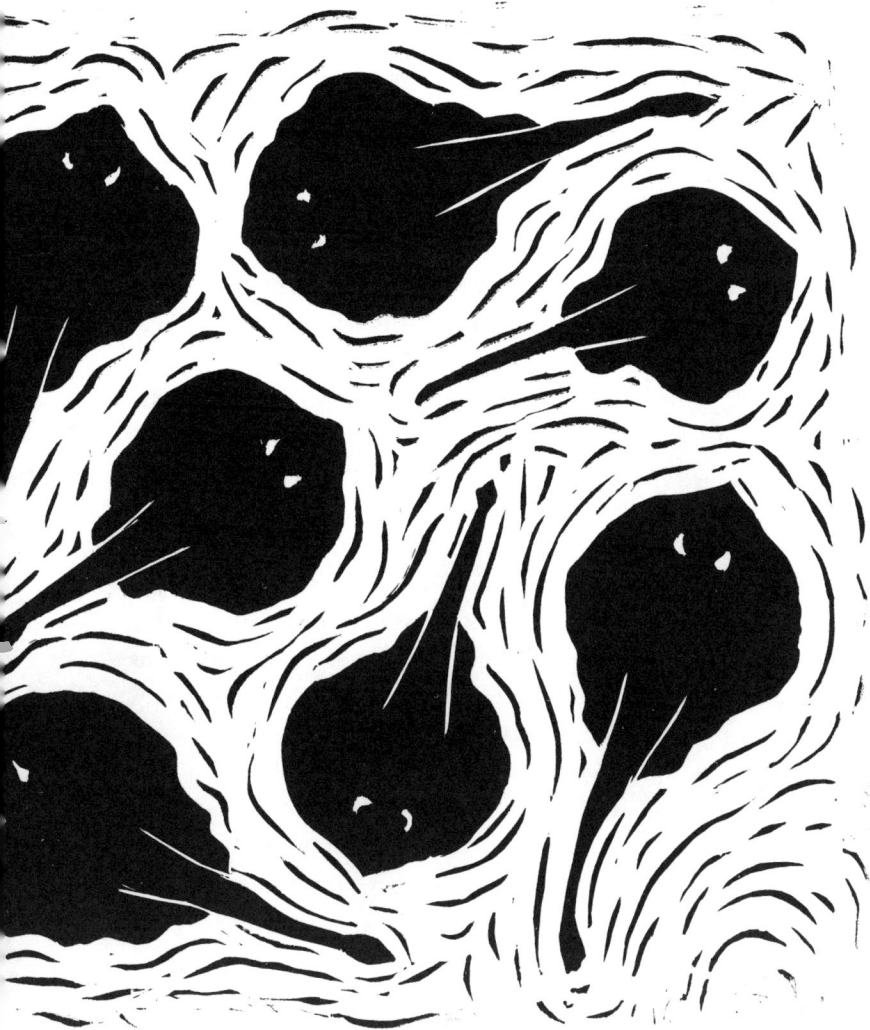

It's not that I disregarded the guidebooks' warnings when I decided to go to Abreojos; I just assumed that I could keep my feet off of the seafloor. I would stay in my space—on my board, on the surface of the water—and they would stay in theirs, on the bottom of the ocean. Stingrays, regardless of their level of notoriety, were no reason to stay away from a world-class surf spot.

After Paolo and I packed a white Ford F-350 with camping gear, surfboards, water jugs, and a cooler full of quesadilla makings, we headed north on Baja's Highway 1. We routinely worked dawn-to-dusk days getting courses to and from their launches and dealing with issues ranging from blown head gaskets to broken bones, so we counted the hours until we could escape our headquarters on the flatwater side and get to the Pacific coast to surf.

Just past the historic town of San Ignacio with its stately old church and dense grove of coconut palms, we turned left onto a vertebrae-rattling dirt road. It ran west for thirty-some washboarded miles before coming to a dead end. In 2005, that dead end held nothing more than a shack selling Tecate, potato chips, and canned beans. We arrived with just enough light to pitch a tent and gawk at the perfectly peeling waves. Their crests must have been a quarter-mile wide, with curls of whitewater folding over on themselves slowly, steadily, and gently, from one end of the bay to the other. The guidebooks weren't kidding; it looked like the perfect longboard break.

As we made dinner, the only other sign of human presence was the light from the little store in the distance. We ate in silence, mesmerized by the sound of waves lapping rhythmically onshore, thinking about what they were going to feel like under our boards.

Stingrays are cartilaginous fishes, which means they are related to sharks. Humans typically have a much greater fear of the bigger, more menacing members of the taxonomic subclass referred to as "the elasmobranchs"—the ones made famous by the *Jaws* movies. This makes sense, since there are very few recorded cases of fatal stingray attacks. The most famous one killed Steve Irwin, Australia's Crocodile Hunter.

Stingray barbs look a lot like feathers but are much more solid. With pointed tips and serrated edges, they could also pass for elongated arrowheads. But they act less like spear points and more like hypodermic needles, hidden inside the creatures' tails. Nestled into grooves on the underside of the barb are venom-filled cells, which burst and release their poison when those tails penetrate foreign flesh.

The poison contains three active compounds: serotonin, 5-nucleotidase, and phosphodiesterase. The first of the three causes intense pain. The other two cause necrosis, or tissue death. Typically, stingrays strike only when threatened by larger predators invading their space. Human body parts that do this are generally in motion, so luckily, most stings to humans are glancing blows.

Steve Irwin wasn't so lucky. A stingray's barb pierced his heart.

As far as I could remember, the only close encounter I'd ever had with a ray was when I visited the Cayman Islands with my family in middle school. On a scuba diving excursion to "Stingray City," the trip leader gave my brother and me little morsels of canned chicken and told us to hold them out while underwater. "They'll come. Don't you worry," he said. I watched as a ray with a three-foot wingspan descended upon my hand, hovering over the chicken chunk momentarily before vacuuming it up with the mouth that sits on the underside of its body. I felt a pleasant tickling sensation and smiled as best as I could with a regulator in my mouth.

Stingrays just didn't seem all that scary to me.

Nevertheless, they were on my mind as I carried my board down to the Abreojos waterline. I paused to listen to the sizzle of wave foam on my feet, once again awed by the jarring interface of contrasting environments that is the Baja peninsula—the land often described as "where the desert meets the sea." Behind me lay miles and miles of columnar cactuses, thorn-encrusted ocotillos, desiccated grasses, and sand. Ahead of me, turquoise, wind-textured salt water spread to the horizon.

I shuffled my way into ten or twelve inches of water, slid atop my board, and paddled out.

When he was killed, Steve Irwin was snorkeling in chest-deep water near Batt Reef off of Australia's northeast coast. He was there to film an episode of *The Ocean's Deadliest*, a documentary collaboration with Phillipe Cousteau, but conditions on September 4, 2006, were too unsettled for the crew to work. Instead, Irwin went out in pursuit of footage for a children's show he was creating. He spotted an eight-foot-wide short-tailed stingray and followed it. According to a cameraman on Irwin's crew, the ray abruptly reared up and struck him in the chest hundreds of times. Irwin was left floating in a cloud of bloody water with a gashed torso, a punctured lung, and a hole in his heart—the hole through which the venom entered. The entirety of the attack was captured on video, as Irwin had instructed his team to record everything that happened in the water, no matter the situation.

The footage was destroyed after the Queensland police watched it and concluded that Irwin was not harassing the ray.

Not actively, anyway. It's unclear how rays perceive our presence, especially when we enter territories that have traditionally been theirs.

After about fifteen minutes in the crystal clear, 68-degree-Fahrenheit water, I got caught up in wave hunting and forgot all about the rays. To this day, sitting on my board and sizing up each and every incoming wave for its surfing potential is the quickest way to quiet the chatter in my busy brain. This single-minded focus on the water's constantly changing colors and textures leaves no room for revisiting the past, predicting the future, or thinking about anything other than whether or not to paddle for the incoming wave. And once I decide to go for it, the world shrinks even further. There is only the interaction between my body and the wave—the feeling of how my weight reacts to the force of the water, how the water reacts to the force of my weight, and how we travel together through a small slice of space and time. This is, for me, a feeling worth chasing.

Toward the end of one particularly long ride on a perfect Abreojos peeler, I stepped a little too far forward on my board, causing the nose to dive. The tail of my board flipped up, and I slipped into the water, which was, at that point, fairly shallow. The instinct to touch down kicked in before I could arrest it, and as soon as my foot hit bottom, a shiver shot up through my core. I hadn't landed on solid ground; I had landed on a layer—or perhaps multiple layers—of fluttering flesh. After the briefest of contact, my foot retracted, as though intuiting that it had crossed a line. My leg jerked upward before I consciously registered the sensation. Several seconds later, with a hefty dose of adrenaline pumping through my veins, I realized that I had just experienced the infamous Abreojos stingrays.

I slid back onto my board and paddled over to Paolo, still panting. "Holy shit, have you touched bottom yet?"

He shook his head no. "I'm trying really hard not to. From the look on your face, I'm guessing you have."

I breathlessly attempted to describe the sensation I had just experienced. "Oh my God. So creepy. It's like . . . it's like standing on a vibrating rug of . . . I don't know what." I shuddered. "Totally unlike anything I've ever felt. Not in a good way."

He squinted. "Are you heading in?"

I had never imagined that the seafloor would be a literal undulating and uninterrupted carpet of toxin-equipped sea creatures. The chances of getting stung suddenly seemed a lot higher. I was rattled, but I thought my close call would keep me honest and make me more aware of my feet. And really, how bad would a sting be? Annoying, at worst, I figured. Besides, I couldn't bear the thought of heading in. "Nah. These waves are too good to turn tail and run. I'll be more careful."

I can't remember how many more ecstatic, effortless rides I got after that. I do remember their consistency. If I sat in the same spot and executed the same series of maneuvers, I could count on a glorious trip across a glassy wave face, during which I could stand up or crouch, walk forward or back, ride down the line or carve. No matter what I did, my mind was blissfully blank.

There's a noticeable lack of data concerning the number of stingray injuries that occur per year in the world. Since stingray interactions are less dramatic than shark attacks and rarely fatal, no one's really keeping track. However, evidence from Southern California suggests that stingray encounters are on the rise there. In the fall of 2017, Huntington Beach lifeguards treated 73 stingray-induced wounds in one day. Lt.

Claude Panis of the Huntington Beach Marine Safety Unit told *National Geographic* that he had never seen even 45 stings in a day during his forty years on the job. Two years later, Huntington Beach smashed that record with 176 stingray injuries.

That rise may have to do with increasing water temperatures, since stingrays love warm water. Of course, human beings have been raising the water temperature all over the globe throughout the course of my lifetime. This temperature change does more than drive the stingrays toward Huntington Beach. Along with the ocean acidification caused by our fossil-fuel consumption, it alters the entire marine nutrient and food web upon which they—and countless other species—depend.

Watching YouTube clips of Steve Irwin, for me, is like surfing an ethical and emotional wave. I can't tear my eyes away as a carpet python bites him on the right cheek. "Son of a gun!" he says, while it continues to encircle his left forearm. I listen enrapt as he describes the bearded dragon's mood swings or plots an intricate crocodile capture designed to minimize the creature's pain. The scenery is captivating, the animals are fascinating, and the passionate lilt of Irwin's Australian accent is both endearing and infectious. "You need to come with me and be there with that animal," he says. "Share my wildlife with me because humans want to save things that they love." I support this idea and know it to be true.

But there are moments when these videos make my skin crawl. "During the dry season, these pools are refuge to a myriad of wildlife, including 'freshies,'" he says in one clip—just before he lowers himself into a desert water hole, submerges, then pops up to the surface clutching a two-foot-long freshwater crocodile. "Have a look at this little boy!" His voice sparkles with admiration and excitement. It's impossible to know how the creature feels about being clasped around the torso by a set of powerful human hands. Reptile faces are so different from ours—their eyes so distant and expressions so alien—that it's inappropriate for me to even hazard a guess. I know how *I* feel though: uncomfortable with the intimacy of this interspecies contact. I don't like seeing him in the pool, and I don't like seeing him touch an animal that should live a life free of human interaction. After a couple of seconds, Irwin releases the wild animal, then turns to his wife and says, "You should come in; the water's so cool." She does, and soon there are two human beings in a roughly ten-foot-diameter water hole in wild, arid, northeastern Australia.

In February 2019, the organization People for the Ethical Treatment of Animals took to Twitter to denounce a Google-Doodle honoring Steve Irwin on what would have been his fifty-seventh birthday. In their words, "A real wildlife expert & someone who respects animals for the individuals they are leaves them to their own business in their natural home." Their tweet relaunched a controversy that had surrounded Irwin throughout his career. When is it appropriate to enter an animal's space? How much encroachment is too much? Is it okay to cross a dividing line if you're doing it in the name of education?

I had just finished another flowy ride and was paddling back out through the whitewater when a bigger set came in. The first of the incoming waves looked like it might break on my head, so I dug in, paddling harder to try to get over it in time. Every surfer knows what happens when you miscalculate this maneuver: You feel the curler lift the front end of your board and flip you over backward. You are simultaneously thrown from your board, spun around, and submerged, then typically thrust into a disorientation that can last anywhere from a couple of seconds to a half a minute. Not fun. During that seeming eternity, you care only about where the surface of the water is and how soon you can get to it. There's no room to think about anything else, such as the presence of stingrays. So without knowing it, I put my foot down, again.

When the barb sank into my toe, the strike itself didn't feel all that different from any other cut. What felt different was the split second before it—the moment when the ocean oscillated beneath my toes. Nausea flooded me, as if some primordial sixth sense knew a rush of toxin was imminent, and my body was trying to preemptively purge it.

The laceration hurt enough that I paddled in—carefully, willing myself not to put my foot down again until I had driven my board up onto the sand and flopped onto the beach like an elephant seal. By that time, it felt like a hot poker had been shoved up through my shinbone. Had I not stripped off my wetsuit immediately upon hitting the sand, I would have needed trauma shears to cut it off. Because the toxin's effect increased with every second, the pain of pulling anything over my skin just a couple of minutes later would have launched me into orbit.

Paolo came in shortly thereafter and found me clutching the truck bed, as though my grip on the metal might discharge the pain. I imagine he simultaneously heard my expletives and saw the blood draining from my right big toe. By the time I had removed my wet bathing suit and crawled naked into my sleeping bag, I was shivering violently. I lost track of the tent, the desert, and the ocean and could focus only on the escalating needlelike nerve pulses in my leg.

Paolo unzipped the bottom part of my sleeping bag and placed my foot in a pot of hot water, the one recommended immediate treatment for stingray wounds. As he clasped my calf to keep it still, the rest of my body thrashed around. I alternately held my breath and panted as my muscles twitched violently—an effect of stingray venom that has been granted the glamorous moniker "facsiculations."

I tried counting in my head, singing Manu Chao songs, and yanking on the drawstring of my sleeping bag's hood. I couldn't stay focused on any of these tasks for more than a couple of seconds, after which another tidal wave of sensation would swell up from within, uproot any sense of physical stability, and toss me back into the inferno. "Don't we have any painkillers?" I cried.

Paolo went to the truck and rifled around in the back seat. "Tequila?"

"Give it to me!"

After he watched me down a double shot in one swig, he remembered that we had run into friends earlier in the day who were camped a little way down the beach. "I'm going to head over there and see if they have anything, okay? I'll be back as fast as I can." He returned with a little orange bottle. "Well, they're painkillers. Juvenile ones, leftover from their kid's recent surgery." I took four.

At some point, Paolo dug around in our first-aid kit and found a Sawyer extractor, a little plastic contraption devised to suck venom from snake bites. "Couldn't hurt, right?" he asked, placing its rubber cup on my toe and pulling back on the syringe-like plunger. The blood seemed to drain from Paolo's face as my own blood accumulated in the tool's reservoir; it was a decidedly unnatural shade of deep purple and had the consistency of Jell-O.

I wondered if I was about to lose part of my toe to a creature I had thought of as relatively harmless—a creature that had once eaten chicken out of my hand.

Then I wondered if I should have been in the water in the first place.

After Steve Irwin died, the Australian government purchased a 330,000-acre former cattle ranch in northeast Australia to honor him. Just a few days after the transaction, plans for the expansion of a nearby mine were announced. Irwin's widow and children spent six years raising the funds and popular support needed to protect the land from this incursion and to formalize its federally protected status. The Steve Irwin Wildlife Reserve is now a "Strategic Environmental Area" administered by the Australia Zoo. It encompasses thirty-five ecosystems and is home to 170 species of birds, 20 species of mammals, and 48 species of fish—including a freshwater cartilaginous fish called the whiptail ray. Apparently, Irwin bought up numerous other tracts of land in Australia, all with the intention of setting aside territories for wildlife to thrive—territories in which human beings are not the dominant residents.

Steve Irwin built his reputation on crossing lines—on entering spaces that few human beings have been capable of or willing to enter, spaces whose borders were drawn largely by the exigencies of human safety. I find it fascinating that his legacy has been to draw more lines, although the boundaries around his reserves are of a different nature. They protect animals and ecosystems from us, not vice versa.

Not only am I ambivalent about Steve Irwin's actions, but I'm also confused about my own. While I taught my students myriad techniques for camping and traveling in the wilderness without adversely impacting the land and its nonhuman inhabitants, it had never really occurred to me that surfing at Abreojos might have been an invasion of the stingrays' space.

I don't think it was as egregious as jumping into a crocodile's watering hole, but I was thrill seeking in an area known to have a concentration of animals that prefer to be left alone. I doubt the presence of two people in the water on a random Tuesday in December had much of an effect on the stingrays; however, two people can turn into twenty and then into two thousand, as recreational sites become better known and people travel more frequently and farther from home to find places to play. There are a lot of other places to surf in the world. Maybe I should have avoided this one.

Similarly, few people are likely to pick up giant pythons as frequently and brazenly as Steve Irwin did, but it's entirely possible that footage of his crocodile-wrestling escapades emboldens Yellowstone National Park visitors to approach grizzly bears or take selfies with bison. These interactions often result in attacks on human beings, and attacks on human beings often result in euthanized wild animals. Drawing lines that prohibit human-animal interactions can be an effective way to limit these tragedies.

At the same time, I wonder if this boundary drawing always serves us. When we leave other creatures alone, we also eliminate the possibility of meaningful interaction between species—the kind that often produces a passion for protection of those species and their ecosystems. Good fences make good neighbors, but they don't always make neighbors that understand and care about each other.

These days, I live on the central coast of California, where I surf with endangered sea otters and dwindling populations of giant and bull kelp. I know that six species of endangered whales might cruise by on the horizon, and a line of endangered brown pelicans might glide overhead as I wait for my wave. I write about and work for the health of these creatures and the Monterey Bay ecosystem they inhabit because I hang out with them on a daily basis. The relationships I have with them were forged by spending time in their space.

My surfer friends estimate that the number of folks in the water has doubled since the pandemic. There are days I would like to go out but don't; there's simply no room for me in the lineup. How do we know when the negative consequences of our presence outweigh the potential benefits of communing with kelp and California sea lions? Should we limit our numbers? These questions nag at me, yet they are dwarfed by the grim reality of climate change: Drawing lines protects animals only from our bodies, not from the pernicious effects of our carbon dioxide production.

In the end, the consequences of my attack were fairly small; I didn't even have to go on the course of antibiotics often necessary to prevent secondary infections after a sting. Microbial hitchhikers can ride in on the stingray's barb, and a variety of seawater bacteria that flush into the wound can cause problems as well. Perhaps the blood purging performed by the Sawyer extractor cleaned me out, because aside from having to scrape off some dead skin and endure the sluggish pace of healing associated with stingray wounds, the little hole on my toe didn't trouble me. My immune system killed off the foreign critters, closed the opening, and tried to go back to being its own isolated universe. Really, the stingray was just trying to do the same thing: keep its own environment clear of things that don't really belong in it.

It's been fifteen years since my stingray encounter, but I still think about it often—especially when there's news of a local shark attack. In May 2020, Santa Cruz surfer and surfboard shaper Ben Kelly was killed by a more infamous member of the elasmobranch subclass: one we call "the great white." The attack took place at Manresa Beach, just ten miles south of the breaks I frequent, as the pelican flies.

Some days, after surfing, I scroll through marine science websites like the one that recently stunned me with its photograph of a stingray skeleton. A researcher had painstakingly reassembled the white cartilaginous "bones" over a background of black velvet. Its architecture was far more delicate and complex than I had expected. What struck me was not the length of its tail or the mean look of the barb's serrations—the ones that had once entered my body. Instead, I was amazed by the structure of the stingray's fins, or, as I often end up calling them, "wings." The specimen in the photo must have had over a hundred rib-like needles extending from its center out to the edges of what was once the rippling flesh that allows it to move through its home environment. The collection of spiny cartilaginous tines looked more like one of those fancy Victorian-era combs than the structural elements of a living, swimming thing. And then—just below its organ cavity and above its tail, in the spot where a human pelvis would be—there was nothing. Just black velvet shining though empty space. Looking at the photograph, I was flooded with awareness of a basic truth: This creature is entirely unlike me.

I stared at the image for a while, glancing away only to make sure that the scar on my right big toe was still visible—as though I needed confirmation that, yes, our paths had crossed once. In that moment, I realized that I was proud of my scar, though not in the way someone might brag about a battle wound. It was a pride borne of connection. I felt lucky to have known stingrays more intimately than the average human being has, to have been forced to consider and care about them.

I was reminded that my daily interactions with ocean-dwelling creatures are precious gifts, no matter how confused I might be about their impacts. These relationships ripple the lines between us, extending my empathy far beyond its obvious borders.

As I looked back at the photo, I could feel a tear forming in my eye—from love, shame, respect, or sadness; I don't know. Perhaps the combination of these emotions created that drop of salt water, our shared space.

LOSING REFUGE

Every winter, thousands of elk inhabit the open fields that make up the National Elk Refuge in Jackson Hole, Wyoming. They are the definition of regal. Their powerful haunches—designed to support five hundred to seven hundred pounds of muscle, bone, and hair—contribute to this impression. So, too, do their necks, which are long, thick, and blanketed by a lush layer of sienna-brown fur. But it's not just their anatomy that grants them elegance. Their slow and steady gait conjures up images of kings and queens in procession, even though they're often trudging through mounds of snow-covered sagebrush. They seem to pause for half a second before lifting a leg, then again for a moment before planting a foot with confidence. Instead of struggling through this uneven terrain, each enormous animal appears to float forward like a hairy cloud until it pauses to lower its mouth to an exposed clump of tall grass.

Outside the National Elk Refuge, cars speed by on Wyoming State Highway 191. Planes take off and land at the ever-expanding airport just five miles away. Tourists buy T-shirts, eat steaks, and learn to two-step in cowboy bars downtown. The incessant human activity—activity that created the need for the refuge in 1912 and has increased exponentially since then—doesn't seem to faze the elk at all. It's as though the craggy peaks on the horizon have granted them dominion over the area, announcing that the animals have always been there and will always be there.

Elk congregate on the refuge between November and March for several reasons. For starters, it's an appealing parcel of land, occupying a broad, flat plain at six thousand feet of elevation. While snow still covers the ground there for most of the winter, it's nowhere near as deep as it is in the mountains. When patches of it begin to melt, the elk can graze on the grasses and shrubs that dominate this treeless landscape. It's much less windy down in the flats than it is up high, and even when the temperatures are well below freezing, the sun is often out, bathing the creatures' dense fur with its gentle warmth. In addition, human habitation has made it impossible for elk to access the grazing grounds their species once used. The refuge is their only adequate substitute. Finally, the spot is made especially attractive by the piles of alfalfa pellets distributed throughout the winter by the US Fish and Wildlife Service, the agency that manages the refuge.

This initiative, called the "supplemental feeding program," is one of the most controversial issues in Jackson Hole—right up there with affordable housing (there isn't any) and the aforementioned airport expansion (there's a lot). It was instituted in 1912, the same year the refuge was founded, with a noble goal: saving the elk. The permanent human settlements in Jackson that started popping up around 1885 had displaced the creatures from their grazing grounds and cut them off from their traditional migratory routes, leaving them stranded and starving. Supplemental nutrition kept them alive, which pleased the tourists and the residents that made a living off of out-of-town visitors. It attracted more animals to the area, which pleased the hunters and hunting outfitters who were allocated more elk tags. It also ended up concentrating elk within the refuge's borders, which pleased the ranchers and farmers who didn't want them interacting with their livestock and eating their hay. For years, feeding these wild animals seemed like a great idea. The

decision to create a structure for the management of another species—seemingly for both its benefit and ours—was lauded as a giant step forward in wildlife conservation. But that was three decades ago.

The first time I saw a herd of elk—not just eight or ten animals but hundreds of them—I was twenty-six years old and instructing a backpacking course in the Absaroka Mountains, about an hour east of Jackson Hole as the crow flies. At the time, I was leading six or seven 30-day expeditions a year, teaching groups of twelve to fifteen young adults how to camp, travel, and get along with each other in the wilderness. The school I worked for ran only a handful of courses in the Absarokas because of the area's preponderance of grizzly bears. Even though bear camping and traveling protocols are notoriously inconvenient (think: hanging food every night, cooking two hundred yards from your sleeping area, and constantly making noise while you hike), I adored the Absarokas for their wide, gravelly rivers and plentiful wild game. We saw bears, bighorn sheep, moose, and elk on every course, so I was happy to work these contracts.

It was a windless late June day when a group of five students and I hiked up to the eleven-thousand-foot high Shoshone Plateau to escape the ubiquitous mosquitoes. We left camp covered in nylon and netting and made our way toward what the map suggested would be a flat meadow the size of fifteen or twenty football fields. As we crested the ridge, the multicolored brilliance of the wildflowers struck us first. But as soon as we lifted our eyes toward the horizon, we found ourselves staring into a sea of brown fur. On the other side of the plateau stood an enormous herd of elk. I could not see individuals well enough to count them, but I guessed they numbered well over five hundred. I was stunned into silence. I gaped as the wind whipped around our hoods, reminding us that we were finally safe from the buzzing madness. "Looks like we're not the only ones trying to escape the insect hordes today," I eventually said, smiling.

"Wow, no kidding," one student said. "I don't think I have ever seen this many animals in one place. Ever." I don't think any of us had. So we plopped down where we were, pulled out our bags of nuts, and passed the binoculars around. I liked looking through the lenses to study a few statuesque individuals as they snacked on the meadow's bounty, but I preferred to just watch the whole throng of them at a distance. From where I sat, they were almost like one organism—a gracious, intentional, and dignified one.

I didn't see the much larger Elk Refuge herd for another couple of years, when I pulled into Jackson Hole in December 1998. I had been distracted by the spectacular views of the Grand Teton and its neighboring jagged summits, as many drivers are when approaching the town from the north. But just before the refuge boundary, I shifted my attention toward the less dramatic eastern side of the valley. There, I spotted a carpet of elk. I veered into one of the pullouts on the side of the road and yanked my binoculars from my glove compartment. For a few minutes, I strained to see antlers, legs, and necks. Then I gave up and, just like I had in the mountains, sat back to take in their plentitude.

They like density, I thought to myself. *There is safety or comfort or success in numbers for them.*

Instructors who worked extended wilderness expeditions liked density too, it seemed. Since we spent a majority of the year camping and traveling with inexperienced sixteen- to twenty-three-year-olds, we were excited to make the most of our months off (December and January) by recreating with other competent outdoorspeople. None of us paid rent anywhere—why bother when you're in the woods for three-quarters of the year?—so we found places to sublet in fun mountain towns. Jackson Hole, with its unparalleled access to backcountry skiing, was one obvious option. But even in the late 1990s, the town was prohibitively expensive for someone on a field instructor's salary. As a result, lots of my coworkers rented houses in Teton Valley, Idaho, a valley just west of Jackson Hole containing three smaller and much more affordable towns. In December 1998, as a twenty-eight-year-old full-time outdoor educator, I shared a three-bedroom cabin in one of those towns—Victor—with four other people. Five years later, I bought a house just a couple of miles away.

The biggest of the three Teton Valley towns, Driggs, had a yoga studio that intrigued me. By my late twenties, I was already experiencing chronic pain in my lower back. I suspected it was from a combination of carrying seventy-five-pound backpacks and sleeping on the ground for much of the year, but I had no intention of ceasing to do either activity. Yoga seemed like it might help me better understand—or perhaps alleviate—my condition, so I bought a ten-punch class card and started to show my face at the studio when I was in town.

Over the course of my first winter in Teton Valley, I found that the days I practiced in the warm, well-lit space were sometimes pain-free. I also felt less goal-oriented on the mat than I did in my "normal" life. Even though I wanted to get better at the harder poses, I was enjoying just playing with them. At first, I was turned off by what I perceived as yoga's "New Age woo-woo factor," but my defenses wore down along with my stiffness. During class, I listened eagerly to stories about Ganesha, Hanuman, Kali, and Durga, fantastical Hindu gods and goddesses whose superpowers I found inspiring. I read about the *yamas* and *niyamas*, ethical precepts for a right livelihood, and they made sense to me. As I let the philosophy in, I was further drawn into both the practice and the community.

The studio owner and all of the teachers who worked for her had received their training through a school called Anusara Yoga. I quickly learned that Anusara was a relatively new tradition founded in 1997 by an American man named John Friend. Friend had come of age teaching Iyengar Yoga, a style that focused on the practice as physical exercise and adhered to strict anatomical alignment principles. After a while, Friend found Iyengar's style a bit too rigid and militaristic for his— and he believed, many Americans'—taste. He embraced the alignment principles and the safety Iyengar Yoga offered, but he wanted to see a little more energy, positivity, and love for life in the practice. So he founded his own school that infused the more optimistic and accepting aspects of Hindu philosophy— such as the idea that we are all fragments of a greater con-sciousness, embodied to experience what life has to offer—into Iyengar's biomechanical principles.

As an athlete, I appreciated Anusara's attention to physical safety and physiological awareness. As a seeker, I appreciated hearing tidbits of wisdom and thinking about their applications to daily life. And like anyone, I enjoyed having an uplifting experience first thing in the morning. But I was apprehensive about getting involved with a yoga "tradition"—a body of knowledge and rituals that bound its practitioners together with a set of common beliefs and actions. *How was this differ-ent from a cult?* I asked myself. *Or the insular and irrational Catholic church I was raised in?* We chanted in Sanskrit at the beginnings of classes; hadn't I done enough of that in Latin already? And wasn't it odd that all the teachers used the same catchphrases and comparisons in their classes?

Still, I understood the benefit of a system of learning and practice. Standardized teaching styles made for consistent messages. Certifications set quality standards and gave practitioners something to strive for. The establishment of a formal community structure made it easier to transmit teachings and bring apprentices up through the ranks. Organization and association were how ideas were disseminated in the modern world. I got that.

So I congregated with other yogis in tight spaces to get healthy servings of what the practice had to offer. There were big benefits to doing so. At the time, I didn't see any real risks.

After having witnessed the sheer magnitude of elk in Jackson Hole during the winter of 1998, I decided I had to go on one of the refuge's infamous sleigh rides. The narrated hour-long excursion from the Visitors' Center out into the middle of the supplemental feeding area is a classic tourist attraction during the ski season. It had always sounded a little cheesy to me—just the term "sleigh ride" called up clichéd Christmas memories that made me cringe—but I knew it was the only way to get a close-up view of the grazing animals.

On the day of my ride, the snow cover was thin. So, rather than being driven in a sleigh with metal runners, we rode in a wheeled wagon pulled by a team of Clydesdale horses.

After about ten minutes of bumping along, we pulled up right alongside a group of about two hundred elk. Unlike my wilderness sightings, when I strained to see individual animals from a distance, I could almost reach out and touch these sturdy ungulates from the back of the wagon. Most of the animals were resting belly-down on the ground—and who could blame them, given their immensity? Up close, I was able to truly appreciate their massive bodies: Torsos on the bigger males were the size of my pickup truck's bed, and their necks were so thick I wouldn't have been able to wrap my arms around them in an embrace, even if they had allowed it. But I wanted to. Their coats looked soft and inviting, and I found myself physically drawn to the steady and peaceful presence they appeared to assume.

I knew that elk were powerful, muscular animals, but so often I had thought of them as larger versions of deer. While elk and deer are both part of the same family, Cervidae—hoofed, vegetarian, antler-growing mammals—overall size is not the only difference between them. Elk prefer mountainous terrain and often migrate seasonally, while their smaller cousins can live just about anywhere, as many suburban gardeners will tell you with great irritation. Elk fur is typically much thicker and coarser than deer fur, and only elk sport that distinctive shaggy neck mane.

Antler size is another difference between the two animals. Elk antlers can weight up to twenty pounds apiece, and by late September each year, they are solid bone. This, in part, explains another of their key distinguishing features: the stoutness of their necks. They are not simply conduits between their heads and bodies; they are body parts in and of themselves—powerful ones, at that. They have to be, because male elk wield their antlers like weapons during the fall rut. Similar-sized individuals will compete for females by first scraping their antlers on the ground while emitting eerily beautiful bugling noises. Then they lock antlers and push each other around in one of the more striking and dangerous displays of dominance in nature.

As the wagon driver explained all of this, he also waxed poetic about the benefits of the refuge's supplemental feeding program, which we had just seen in effect. It was, in essence, what allowed us to see so many animals so close together and so close to the wagon's path, since they were gathered around the piles of food, doing what they needed to do to survive. I don't remember him saying anything negative about the program or the gatherings of elk it fostered. It was the late 1990s, a time that now feels innocent and distant, a time before we were thinking about the long-term consequences of concentrating animals in small artificial habitats.

After about ten minutes of observing the first group, we moved onto another and watched them casually eat the hay and alfalfa pellets that had been delivered by refuge staff earlier that morning. The wagon driver had told us that, after the mating season, elk typically split into bull herds and cow/calf herds. I couldn't see the boundaries between these groups from my hard wooden bench seat, but I thought that perhaps the herds intermingled for parts of the day. Or maybe the artificial concentration of elk brought about by the food delivery was blurring the borders between the groups. I had to wonder if our interference with their feeding behavior was altering not just their migratory habits but their social structure as well. But it was for the better; the driver had told us that numbers had been steadily increasing. More animals is always a good thing, right?

I definitely thought so at the time. As someone who has always worried about the planet's imbalanced ratio of human beings to other living things, I found the sight of an enormous elk herd reassuring. I hadn't yet considered how concentrations of animals—or people—could cause problems.

The first John Friend workshop I attended was held in a former gymnasium in Park City, Utah. The space had once been part of a high school, but a wealthy Anusara Yoga practitioner had purchased the high-ceilinged, wooden-floored building and converted it to an enormous studio complete with two- and three-foot-high brass *murtis*—statues of Hindu deities—lining the stage. About 250 of us laid our yoga mats in even rows and waited for Friend to stroll to the front of the room, turn on his headset microphone, and invite us into the practice with his big smile and even bigger presence.

He was an excellent teacher, but what really amazed me was the way he conducted the energy of the room. He was just one barefoot man—and a fairly short one at that—walking the aisles created by our mats. Yet, through some combination of words, emotions, and intentions, he managed to bring 250 disparate bodies, minds, and souls together to breathe, move, sweat, and celebrate their existence in unison. Everyone was onboard; there was nowhere else to be.

Not only was I impressed, but I felt great when I left. The demons of inadequacy that had hounded me since childhood seemed to take a break when I was fully engaged in a strong practice. My chatty inner voices went silent, and I could almost believe that I really was a "spark of divinity," as Friend had said—and that, in fact, we all were. Whatever this was, I needed more of it in my life.

It was hard to come by though, since at the time, I was spending large portions of the year teaching and administering wilderness courses in Mexico and Chile. And I was getting tired of it. I had been working for the same outfit for twelve years, constantly moving around for most of that time. Even though I'd purchased a house in Teton Valley by then, I was still cooking and sleeping in tiny communal living quarters in other countries. I had made new friends in Idaho, including people that weren't in the outdoor industry and didn't under-stand my multi-month absences. I wanted to ski more, bike more, and do yoga more, and I was beginning to think that having a less exotic but more routine job—catering or working at a coffeehouse, for instance—had a lot of appeal.

One of those new Idaho friends was a woman named Cate who was a teacher at the yoga studio. She was a certified Anusara instructor, which meant that she had garnered the prestigious credential—referred to by some as a "Yogi PhD"—that put her in John Friend's inner circle. It enabled her to both lead teacher trainings and bestow credit hours on students working up to those teacher trainings. I envied her role.

One day in February 2007, I was in Chilean Patagonia wearing a hat and a down jacket in the office despite it being the middle of the Southern Hemisphere's summer. I was a thirty-seven-year-old so fed up with communal living that I had almost convinced myself to hitchhike alone on the Carretera Austral—southern Patagonia's sole (unpaved) highway—to escape the tiny farm town where our school's operations were based.

It was at that moment that Cate emailed me and asked me to get in touch. Instead of watching one of my rationed *Gray's Anatomy* episodes on DVD that night, I went back to the office to receive an international phone call. I listened as Cate explained that the yoga studio in which we had both practiced for years was up for sale. The owner and principal teacher had been injured in a bus accident and wanted to focus on her rehabilitation. Cate was the obvious choice to take the helm, but since she had another business that consumed her time and provided her with a handsome income, she needed a partner who could devote energy to the day-to-day management of the place. That led her to me.

Her offer felt like a lifeline. I grabbed it and yanked myself—for once and for all—out of the peripatetic outdoor instruction life. I could live in my house full-time and have a meaningful job. I could start teaching yoga and work toward certification. And I could provide other people with the physical healing and greater emotional peace yoga had provided for me.

A month later, in May 2007, I was back in Teton Valley, Idaho, as the co-owner of an Anusara Yoga studio and the twenty-five-hundred-square-foot building in which it was housed. How I thought I was going to make a living running this place and teaching classes when the total population of the surrounding community was six thousand people, I don't know. I do know that, at the time, it seemed like a giant step forward for my health and happiness. So I made my choice.

Prior to about 1884, there were no permanent settlers in Jackson Hole, and it's estimated that between twenty thousand and thirty thousand elk wintered in the area. But after people took up residence in this challenging ecosystem and began fencing in land to raise cattle and grow hay, elk numbers began to drop. Then, three very cold winters—1909, 1910, and 1911—caused those numbers to plummet. According to Jackson Hole Historical Society records, elk wandered the streets during the winter of 1909 and died of starvation in the middle of town. "One resident noted . . . that it was possible to walk at least two miles stepping on elk carcasses without ever putting a foot on the ground," one particularly outstanding sentence claims.

Two groups of people were especially worried about this trend: villagers who were concerned about the elk's demise and ranchers who were angry that hungry elk were raiding their feed stores. They came together to purchase hay for distribution to the starving elk and successfully petitioned the state of Wyoming to follow suit. S. N. Leek, a local hunting guide, took photos of the tragedy with a camera given to him by a client, George Eastman of Eastman Kodak. Leek's publication of these images, along with the articles that accompanied them and the lecture tour that ensued, drummed up national concern for the animals' plight. There was enough concern that in 1911, the federal government allocated twenty thousand dollars for feeding the elk. In 1912, another forty-five thousand dollars was used to purchase 2,760 acres of land. This step created the

National Elk Refuge. In 1927, 1,760 more acres were added to the parcel, then 17,000 more in 1935. Today, the refuge covers 25,000 acres. This may seem like a lot, but the elk's Jackson Hole range prior to human settlement was about 100,000 acres.

The establishment of the refuge effectively stopped the elk population's downward trend, so increasing its acreage made sense. An increase in the supplemental feeding program had to accompany this growth, since the land couldn't sustain a large herd through the winter on its own. As expected, with more land and more food, more elk congregated on the refuge, and by 1985, refuge staff counted eighty-five hundred animals. To many, that seemed like a cause for celebration.

Yet, by 1985, at least a few people were starting to foresee a new set of issues that would imperil the Jackson Hole elk herd. That year, Refuge Manager Rees Madsen, in his US Fish and Wildlife–published article, "History of Supplemental Feeding at the National Elk Refuge," said, "As these supplemental programs maintain population levels above the carrying capacity of the habitat, related problems are almost certain to develop. Damage to the range from concentrations of animals and disease transmission related to the overcrowding are two examples." He didn't specify which diseases he had in mind. It's obvious, however, that he thought the risks were serious enough to warrant consideration of abandoning the supplemental feeding program, since he went on to say, "The emotional attraction and attachment to wildlife by the public coupled with the resistance to change we all seem to exhibit makes the cancellation of a winter feeding program almost impossible. . . . We [are likely to] find ourselves locked into just such a program because of well-meaning policies developed years ago in response to desperate circumstances."

These two sentences perfectly summarize the predicament the refuge would find itself in just a couple of decades later:

how to navigate the disorienting space between "desperate circumstances" and "well-meaning policies." I would eventually find myself in a similar position.

For a number of years, the yoga studio gave me a reason to wake up in the morning. Yes, I still had to work a bunch of other jobs to make ends meet—I made energy bars, taught Spanish classes, catered high-end weddings, and even did some landscaping—but they were all in service of something greater.

I had fully bought into Anusara Yoga. I was spending hundreds of dollars traveling to workshops in San Diego, Tucson, and Los Angeles. Some of these were with John Friend; others were with his principal teachers—the ten or so instructors who were considered to be his most advanced students and practitioners. Every time I went away for a weekend of immersion, I came back with new knowledge and inspiration to share with the students in my own community, allowing me to justify the investment of time and money. Was I overlooking some red flags? Maybe. But this question is much more obvious to me in retrospect. At the time, I told myself that this was the way the organization worked: There had to be someone at the top, then a tier below, then another below that. And we needed to spend time together to grow the lineage and spread the love.

I also had a goal: certification. Certification would enable me to teach workshops in yogically underserved intermountain west towns—places no one else wanted to travel to that had a hunger for what the Anusara system had to offer. Once I could do that, I could quit all of my side jobs and have one important and meaningful career, like my father and brother and former college classmates had.

That goal motivated me through several years of steady teaching, study, and practice. Finally, in late 2010, the class video DVD I sent to the evaluation committee was deemed certification-worthy. I opened a cardboard-stiffened manila envelope to find a beautiful certificate emblazoned with the Anusara Yoga logo and signed by John Friend himself. The gold seal said that I was Certified Instructor Number 352. I still have the photo my boyfriend at the time took of me the day it arrived in the mail. I am grinning from ear to ear.

I did quit my side jobs, and I began teaching weekend workshops in places like Livingston, Montana, and Pocatello, Idaho. I had it pretty good there for a while, cycling through communities that ringed the Greater Yellowstone ecosystem. Like Jackson Hole's elk herd, I was moving around seasonally. As far as I could tell, it was working for me.

A disease did come to haunt the National Elk Refuge shortly after Refuge Manager Marsden made his ominous prediction in 1985—a neurological illness called chronic wasting disease, or CWD for short.

Because CWD affects only cervids—the class of animals that includes deer, elk, and moose—it is not typically a household term in the United States. However, most people have at least heard of CWD's grisly cousin, mad cow disease. Both conditions are caused by prions, poorly understood pathogenic agents that appear to wreak havoc in the brain by forcing certain proteins to fold abnormally. This, in turn, causes the degeneration of neurons and "spongiform changes." Spongiform changes are ugly. They include stumbling, drooling, listlessness, a lack of fear of people, and, of course, weight loss—wasting. There is no cure for CWD, and it always kills its victims.

In addition to being incredibly destructive, CWD is also highly contagious. It is typically transmitted through bodily fluids—feces, saliva, urine—the kind of thing that accumulates when animals are living on top of one another, like elk on a feedground. Even after individuals die, prions can be shed into the surrounding area and spread to others. It's possible that they can even be stored in the soil where the elk gather, facilitating speculation about "hotspots" and "superspreader events"—terms we humans have grown quite used to, thanks to our own recent global contagious disease outbreak. So far, there have been no recorded cases of CWD transmission from elk to human beings, but the Centers for Disease Control still warns hunters not to eat meat from animals that have tested positive for the disease.

CWD was first identified in captive Colorado mule deer during the late 1960s. However, it was not discovered in a wild animal—and therefore was not much of a wildlife management concern—until 1981. In 1985, the year Marsden suggested that disease transmission could be a risk for elk in densely populated habitats, CWD was found in mule deer in Wyoming. By the mid-1990s, the disease had been detected in captive cervids in Oklahoma, Nebraska, and Saskatchewan, and by the early 2000s, states with recorded cases of CWD included Pennsylvania, Virginia, Iowa, Illinois, Minnesota, Wisconsin, North Dakota, Kansas, New Mexico, and Utah.

Despite Marsden's warnings, the supplemental feeding program on the Jackson Hole Elk Refuge continued even after cases of CWD were reported in wild Wyoming elk. It came under more and more scrutiny every year for the way it fostered ideal disease transmission conditions, but nothing changed.

In 2008, by the time about fifteen states had reported cases of CWD, the refuge committed to a "step-down" policy that stipulated a gradual decrease in supplemental feeding. The policy's intended goal was to bring the size of the herd down to five thousand animals. As a result of disagreements between state and federal agencies, however, neither decrease happened. It had started to look as though the managers, the elk, and the land were all locked into the path that a small group of people had chosen a hundred years earlier.

Alfalfa pellets were regularly distributed for the next ten years, and the refuge's tactic for managing CWD was to "watch and wait." They tested all carcasses taken by hunters and killed any animals exhibiting symptoms. However, an elk can be infected for eighteen to twenty-four months before the stumbling and weight loss begin. And animals can live for up to two years with symptoms before they die. Given the migratory and social nature of elk, it's easy to see the problem this presents.

In 2017, the communities of Star Valley and Pinedale, forty and seventy-five miles from Jackson Hole, respectively, reported cases of CWD in mule deer. And in November 2018, an adult male mule deer that had been hit by a car in Grand Teton National Park—the parcel of public land immediately adjacent to the refuge—tested positive.

In March 2019, as cases crept closer, the nonprofit legal foundation Earthjustice filed a lawsuit against the US Fish and Wildlife Service demanding an immediate end to the supplemental feeding program. In its ruling, the DC Circuit Court of Appeals mandated that the federal agency reduce the size of the herd by taking this very step. But again, nothing happened. In February 2020, Earthjustice filed another suit.

Meanwhile, CWD spread even farther. It has now been found in twenty-nine states as well as in Canada, Norway, Sweden, and Finland. Nevertheless, Wyoming continues to be the condition's ground zero. The Centers for Disease Control's website has a map of CWD's US range—one in which a county gets colored in red when if it's had a recorded case. Most western states have patchwork patterns. Wyoming is a nearly solid red rectangle.

In February 2012, I woke up to a flurry of emails, most of which said something like, "Have you seen jfexposed.com? Holy $@&#! We're done."

"Jfexposed" was a website mounted by one of John Friend's employees with the obvious intention of taking down his boss. It was populated with allegations of tinkering with employee pensions and forcing employees to receive packages of marijuana in the mail. These actions were immoral and illegal but not all that scandalous. What caused the biggest stir were postings of sexually explicit Skype chats between Friend and a married Anusara Yoga instructor as well as the revelation that Friend had been leading a "Wiccan sex coven" he called the Blazing Solar Flame. Several of the women in the group were married, and a few were Anusara Yoga teachers.

In less than twenty-four hours, Friend's lawyers had the site taken down, so I never got to see and evaluate the evidence firsthand. By the time I had the chance to type "jfexposed. com" into my browser window, I got a 404 error. Not that it mattered; the damage had already been done.

The most experienced Anusara instructors, the inner circle, jumped ship first. They decried Friend's unethical behavior, pointing out that he had routinely spoken about the importance of "right action"—impeccable and consistent ethics—in relationships during his workshops, all the while having affairs with married women. They called attention to the inherent power differential that exists between teachers and students—one that Friend had often talked about in teacher trainings—and accused him of shamelessly abusing it. It seemed the farther up the hierarchy the instructor was, the more vociferous they were in their statements. It was as if they believed they could escape the contagion by moving quickly— as if the farther and faster they ran, the more likely they could remain untouched by a disease that threatened to take out their livelihood.

Of course, these teachers also had significant followings of their own. I suspect that more than one of them were looking for a graceful way to get out from under Friend's suffocating thumb. Rumors had begun to circulate about Friend asking for royalties from other instructors' media sales, and some instructors had come under fire for "doing their own thing." Instead of having to challenge their leader by locking antlers, they could use his actions as a pretext to walk away and start their own herds.

One by one, nearly all "big name" Anusara teachers and studios disassociated from the tradition. What remained was a scattering of open-mouthed people staring into an empty feedlot. I was one of those people.

In December 2020, the first case of CWD was found in a cow elk harvested by a hunter in Grand Teton National Park, mere miles from the refuge. It was as this point that it became apparent even to nonscientists that the very structure set up to preserve the Jackson Hole elk herd might contribute to its destruction—and potentially take out other North American herds in the process.

However, CWD wasn't and isn't the only contagious disease in play among Rocky Mountain wildlife. Both elk and bison can contract brucellosis, a disease caused by the bacteria *Brucellosis abortus* that induces abortions in pregnant females. This condition was first found in domestic cattle, and it was domestic cattle that carried the bacteria to Jackson and transmitted it to elk and bison in the first place. Over the years, state and federal agencies have spent billions of dollars working to eliminate brucellosis from the livestock industry, and they have succeeded in doing so on all farms except those within the

Greater Yellowstone Ecosystem (northwest Wyoming, eastern Idaho, and southwestern Montana). In this area, elk and bison have transmitted the bacteria back to cattle, who also abort their fetuses upon becoming infected.

People can get brucellosis, typically from eating unpasteurized dairy products produced by infected animals. We can't contract the disease from cooked meat; however, hunters are still advised to wear gloves when handling elk and bison entrails, since they can be infected with the bacteria from handling contaminated reproductive organs. Brucellosis in humans, which is associated with flu-like symptoms, can be treated with antibiotics.

Neither human nor elk populations have been significantly affected by this disease. Really, the main impact of brucellosis is financial. When it's detected in their herds, ranchers lose baby cows and are forced to test and quarantine their animals, a process that can be both expensive and time-consuming. Brucellosis has put livestock operations out of business.

For this reason, ranchers are more than a little sensitive about the prospect of elk wandering onto their land. In their eyes, the refuge, along with its supplemental feeding program, is the best way to prevent these unwanted interactions. But of course, the refuge and its supplemental feeding program are something like a ticking time bomb for the elk themselves.

The question of how to move forward hangs over the land and its inhabitants.

Anyone who had been exposed to John Friend was contaminated by what was soon being called "The Anusara Scandal." That included me. In a matter of weeks, demand for my workshops evaporated. My spring weekends in Boise and Bozeman were canceled. I didn't have the heart to schedule work for the fall.

Personally, I hadn't been all that surprised by the revelations of Friend's behavior. It was impossible not to notice the amount of attention attractive women showered on him at every Anusara event, and due to the nature of the touring workshop scene, he spent an awful lot of time with them. It would have taken an extraordinary amount of willpower not to get involved with at least a few of the unattached yoginis, power differential or not—and I was under no impression that he had that kind of willpower.

But a Wiccan sex cult? How many famous spiritual leaders had been tainted by scandals of this sort already? I could think of at least four recent international cases—Amrit Desai, Yogi Bhajan, Swami Muktananda, and Khausthub Desikachar. And I had heard about a number of studios that experienced management shifts as a result of owners' or teachers' sexual misdoings. It had practically become a stereotype of the industry, and it embarrassed me. I was angry at Friend for not walking his talk, yes; but I was cynical enough to believe that nearly all people in positions of power did that. I couldn't believe that he had been so careless with his authority, so reckless with a structure that many of us depended on for our livelihood.

What upset me most of all, though, was how he reacted to the outcry. He could have admitted to wrongdoing right away. He could have apologized. He could have turned to the other top-tier instructors and asked their advice, potentially forming a less hierarchical and more collaborative leadership structure. But what he did was go radio silent, claiming that his lawyers advised him not to comment on the matter. I was not impressed.

But I was lost. Although my regular weekly students were still coming to class, I wasn't sure what to do with them. Should I still teach Friend's trademarked Universal Principles of Alignment, even though there was no tradition to back them?

Should I still open classes with the chant that Friend and the world-famous yogi musician Krishna Das had cowritten for us, even though Friend himself had gone underground? Knowing my community had splintered left me listless. I would never get to attend the Certified Instructors' Gathering I had worked so hard to be a part of. I would never again get a big sweaty hug from anyone in the lobby or scan a huge practice room and recognize more than half of the smiling faces there. Most significantly, I would never again look into the eyes of the person on the mat next to me and think, *I see the beauty in them, and I know they see the beauty in me too.*

The spot at the front of the room started to feel very lonely.

One day, as my students were lying in savasana—the final pose of every yoga class in which you rest on your back with your eyes closed to let the experience of the practice marinate in your body—I looked out at the empty field across the street from our studio. I wondered if it had once had elk in it, years ago, before the town popped up, before the wastewater treatment plant took over the middle of the wetlands, before the highway created an obstacle to movement and migration. I remembered that once, an instructor I had worked a canyoneering course with told me that he thought the animal I most resembled was the elk. He had heard that on a previous expedition, I had hiked twenty straight hours to initiate the evacuation of an injured student. This was before the days of satellite phones, so I'd had no choice. He noted that I never moved very quickly but that I had remarkable stamina and staying power, that I was sturdy and reliable, and that I always carried a heavy pack without complaining.

I suppose that was true. I belonged in the mountains at the time; after all, I was tromping through them with young adults for seven or eight months per year. I was sleeping in their shadows at night and exploring their valleys and ridges during the day. Perhaps the rugged landscape sculpted the elk and me in similar ways, in a case of single-generation convergent evolution.

But when the shit hit the fan in yoga world, I felt nothing like an elk. All I wanted to do was turn tail and run.

If packing elk into a concentrated area by feeding them creates, as the DC Circuit Court stated in the Earthjustice case ruling, "a miasmic zone of life-threatening diseases," why are alfalfa pellets still being distributed around the refuge throughout the winter?

For one, many people believe that not feeding the elk will result in widespread starvation. The animals have adapted to a migratory cycle that includes wintering on the refuge, an area that simply does not have enough natural forage to support the current population. If we don't continue to feed the elk who have grown dependent on this source of nutrition, the argument holds, they will die—not in a year or two as they would from CWD, but immediately. No matter that the population size is overblown; starving elk don't make for a good visual in a town that prides itself on rugged physical beauty. And it's hard to explain to visitors and residents alike that elk are being left to die in the present to avoid potential deaths in the future. As a species, we humans don't like to play the long game. We respond emotionally to tragedies that are right in front of us. Spreading alfalfa pellets enables us to avoid immediate tragedy.

Then there's the fact that the supplemental feeding program effectively keeps elk off of local ranches. Ranchers worry about elk eating the hay meant to fatten up their cattle, and they don't like wild animals of any kind roaming through their property. And of course, there's the ever-looming threat of brucellosis. Recent testing of harvested elk in the Greater Yellowstone Ecosystem suggests that their brucellosis infection rate is about 15 percent. While that number is not all that high, it's high enough to strike fear into the hearts of ranchers who have seen their neighbors' herds and yearly incomes wiped out by just a few cases of the disease. That fear prompts these typically longtime and politically influential residents to advocate for the continued isolation and concentration of elk on the Refuge.

Meanwhile, however, this same isolation and concentration of elk promote the spread of CWD. The refuge is left walking a thin line between two diseases: one that can take out families' livelihood and one that can take out an entire species.

Because both diseases are transmitted through close contact, spreading the animals out in areas far from cattle would be an obvious solution. However, human beings live all over the elk's former range; that's why the refuge was established in the first place. If they were to be spread out now, they would likely die from lack of food, lack of habitat, bullets, and vehicle impacts. But if we continue to concentrate them, they may die from CWD, some evolving variant of brucellosis, and the next series of contagious diseases coming down the pike.

This is the classic catch-22 of the Anthropocene: We create a well-intentioned structure to try to improve the lives of another species forced to live with our impacts, and that structure comes back to haunt us.

After the maelstrom surrounding the Anusara meltdown had settled a bit, I could finally lift my head to look around and take stock of my situation. I could try to rebrand myself using some hybrid combination of yoga tradition names to wrap my offerings up in a new and enticing package; plenty of other teachers were attempting this strategy. But as an instructor in a rural area, I had benefited from the pyramid-scheme-like structure that had required students to amass credit hours with certified teachers. Without that structure in place, I would have to rely on my own personal charisma and marketing abilities, and I knew darned well that I lacked sufficient quantities of both to make it. I could go back to teaching only local yoga classes, but that would necessitate taking on multiple side jobs to make ends meet once again. At the age of forty-three, I found that prospect dismal.

Furthermore, I thought it might be time to leave Teton Valley. I loved the landscape and the herd I had come to run with there. And yet, they had both grown too familiar. I couldn't go to the grocery store without hearing (or worse, spreading) some piece of gossip. I had dated all of the single men my age who weren't alcoholics or permanent adolescents—and I had dated some of them too. And I had grown weary of a winter that started with the first snowfall in late September and ended with the last one—if we were lucky—in late May. I started to think I needed a new field to graze in.

More significantly, though, I felt like I couldn't redefine myself. The community had decided who I was: I was an elk. I was strong and stately, and I had the stamina to hold down a yoga studio through an economic downturn in a small town. I was the kind of person who stayed the course and did it with grace and quiet strength.

Only, I wasn't anymore.

Cate and I tried to sell the studio for almost a year. Then we took down the sign and gave away a closetful of mats, blocks, and blankets. Cate bought my half of the building for significantly less than I'd paid for it eight years earlier, and her husband turned the sunny, oak-floored space into a commercial glass shop. Shortly after that, I put my house on the market. A single man bought it a week after it was listed, for the same price I had paid back in 2004.

Many people would say—and did say—that extracting myself from my past choices resulted in enormous losses. In a certain light, I can see that. I can dwell on lost money, lost years, lost direction, and, worst of all, lost community. Anusara Yoga had not only given me a structure and a purpose; it had given me a sense of place among a group of people I cared about and connected with. Those people still existed, but without the organization, our interactions could never be the same. But I rarely let myself think about those losses; it simply hurts too much.

It's much healthier for me to focus on the relief I felt when I drove out of town with a half-filled U-Haul box truck and an open mind.

Much to everyone's amazement, as of the fall of 2024, CWD had yet to show up within the boundaries of the National Elk Refuge. However, most authorities still agree that it's just a matter of time before it does.

During the long, cold, and snow-heavy winter of 2022–23, here were seventy-five hundred elk on the refuge. Supplemental feeding began on January 11—two weeks earlier than the average seasonal start date, even though the US Fish and Wildlife Service's step-down plan stipulates that feeding be started two weeks *later* every year. According the refuge's chief biologist, this choice was made "to prevent large numbers of elk from leaving the Refuge for private land [to the] west and northwest." Apparently, on January 10, staff had watched a hundred animals move into the Gros Ventre River bottom—an area that, before the influx of permanent human residents in the valley, would have been part of their winter range. There are ranches there, and I can only guess at the amount of pressure that was put on refuge staff to keep the elk from moving toward them. And then there are the emotionally laden editorials that appear in the media year after year, exhorting the refuge to prevent the elk from cruel deaths by starvation.

I don't envy these decision-makers.

The big step that was taken during the winter of 2022–23 was the purchase of an enormous incinerator, to the tune of five hundred thousand dollars. This industrial-looking collection of metal barrels on a trailer burns dead elk at 1,800 degrees Fahrenheit, the temperature at which the prions that cause CWD are destroyed. If all dead cervids are burned, the logic goes, animals with undetected CWD can't spread the disease. But of course, elk can live for up to two years with CWD, so by the time these hypothetical animals are incinerated, the damage has most likely been done. And as of yet, the incinerator can be used only for refuge animals, not for roadkill or animals who have died on other nearby parcels of land. The political structures in place seem to be preventing the kind of cooperation that could maximize the effectiveness of this tool, which, I fear, might just be a political move in and of itself.

Supplemental feeding ran until April 8, 2023. Alfalfa pellets were delivered for just shy of three months. Estimates put the 2023 herd at 7,410 animals.

I live alone now, in a California city I moved to in my mid-forties without knowing a soul. I work as an editor, so I spend a lot of time by myself. While I can say that I am loosely part of several communities—a swim team, the regular low-tide surf crowd, the jam band fans, the arts scene—I have not, and probably will not, come anywhere close to feeling the sense of belonging that I once experienced both in Teton Valley and as a part of the Anusara "tribe." Sometimes, this makes my heart ache. Other times, I breathe a sigh of relief for having escaped from a group that was feeling more cultlike all the time and a town where I was feeling hemmed—and snowed—in. Most days, though, I just shrug and give thanks for the beauty of the Monterey Bay that I live along, the amazing diversity of marine and terrestrial life I interact with every day, and the freedom I have to wander anonymously within groups of my own species.

Regardless of how I might feel at any given moment, I recognize how lucky I was to have had the luxury of picking up and moving to an entirely different ecosystem. Sure, there were adjustments; I sold my ski gear, pared down my warm clothing collection, and invested in a second pair of flip-flops. I had to get a new driver's license and find a new health insurance carrier, and a smog test was required to bring my car into a state with stricter emissions laws. I work much harder to pay a much higher monthly rent, and I deal with significantly heavier traffic. But that's about it. I went from living in the Greater Yellowstone Ecosystem to the central coast of California with minimal disruption to my daily habits and minimal effect on my physical well-being.

Elk can't do that. Their bodies evolved in alpine ecosystems. Everything about them, from their fur and their digestive system to their social structure, mating strategy, and migratory pattern, revolves around life in the Rocky Mountain West. They couldn't walk to Santa Cruz and decide to live off of live oak acorns and marine crustaceans. And yet, they might not be able to stay where they are, packed tightly onto the National Elk Refuge and the other twenty-two feedgrounds in Wyoming. They can't go up into the mountains during the winter; life is too strenuous there, and there isn't enough food. They can't be relocated because there's no similar terrain without people and cows. They can't spread out through the high-altitude valleys because people and cows live there. Besides, they're herd animals; they don't want to spread out. They evolved to live together in groups. And the diseases that threaten them evolved to take advantage of that. So here we are.

I tend to jump to the conclusion that we all make bad choices when we're backed into corners, when we feel threatened, when the options appear to be limited. As a result, it's easy for me to see this situation as hopeless. But then I remember that life surprises me all the time. Events surprise me, facts surprise me, and most of all, nature surprises me. We didn't foresee the effects of confining elk to a refuge. I didn't foresee the effects of investing in a highly combustible yoga tradition. John Friend didn't foresee the effects of his extracurricular activities. There is just so much we miss as we navigate our days. The complexity of social structures, ecological webs, and decision trees seems to far exceed our capacity to process them.

And CWD still hasn't been detected on the refuge. I managed to build a new life that challenges me in different and interesting ways. Even John Friend has bounced back; he actually started another new yoga tradition. I have no interest in tapping into it, but last I heard, he's out there doing his thing.

After years of following news about the National Elk Refuge, I know where I stand on supplemental feeding. There are one billion cows in the world and only two million elk. Cows are not endemic to the Rocky Mountain West. Elk are, and preserving the biodiversity of this precious planet is my number-one concern. But I am not a rancher. I don't eat meat, and unless I know my yogurt is coming from happy, grass-fed, hyperlocal animals, I don't consume cow dairy either. This makes it easier for me to land firmly on the side of ending the program entirely, especially since many scientists believe that the herd would not suddenly starve, but that it would instead shrink to a sustainable size over time. They are quick to remind us, too, that a healthy number of natural elk predators—wolves and grizzly bears—would cull the weak and diseased animals, keeping the herd strong. If I had a huge chunk of money, I would start buying up ranches in Jackson Hole and expanding the refuge to help this population stabilization along. But I don't.

And that's okay, because, really, a few wealthy people can't solve this problem. It's going to take a change of mindset—one that begins with an acknowledgment of just how little we know, how little we control, and how poorly we predict the effects of our actions. With humility in the face of mind-blowing complexity, perhaps we can make more nimble choices and hold them more loosely. If we can change courses, adjust strategies, and let go of emotional attachments, perhaps we can react to the ever-surprising course of events with more grace.

First, however, we must remember that we're all in this together, that we are all entwined.

After a few years of living in California, I got involved with a man who lives in Jackson Hole. I met him when I went back to visit friends for what was supposed to be a one-week trip. I stayed for a month; then I kept going back. These visits gave me the opportunity to reconnect with friends in Teton Valley and, slowly but surely, reintroduce myself to this community as a different version of the person I was when I lived there. They also helped me see that the yoga career I dismissed as a failure for so long may not have been as much as a dead end as I once thought. It taught me to see points of intersection, foster connections, and tell stories that braid seemingly disparate threads into organic tapestries. It showed me what it was like to live within the nurturing structure of a herd as well as how to survive without it. And it pushed me to recognize that we can't always see where our decisions will lead, and as a result, we should be willing to move on from them when the time comes to do so.

My frequent visits to Jackson Hole also kept me interacting with elk—both on the refuge and in the wild. On one sunny September day, my boyfriend and I decided to hike up to a lake in Grand Teton National Park that is known for its glacially fed turquoise water and the way it perfectly reflects the biggest mountain in the area, the Grand Teton. There is no trail to the lake, so for many years, its existence was something of a locals' secret. Neither of us had been there for a long time, so we were shocked and saddened when we veered off the main thoroughfare to discover that, in the absence of a Park Service–built trail, numerous eroding user trails had been created. It seemed like everywhere we looked, we saw another patch of alpine meadow scraped bare by human feet. Later, a friend told us that the destination had become an Instagram sensation; the secret was out. The lake itself was still gorgeous, however, and we must have spent an hour lying on a smooth granite rock admiring the mirror image of the surrounding mountains dancing on the water. Then we headed downhill, chatting for a while about the impacts of the park's choice not to build a sanctioned pathway there. After getting through a few steep sections, we hit the flatter portion of the trail and started running. It was late in the day, and our route through the valley had us weaving through foliage in the midst of its autumn color change—mountain ash turning hunting vest–orange, black hawthorn growing red at the tips of its leaves, willows forming yellow canopies by the sides of streams. We had fallen silent and were left listening to only our breaths and footfalls. Then, suddenly, the air was pierced with a haunting cry. We both stopped in our tracks, instantly recognizing the unmistakable sound of a bugling elk.

If I didn't know better, I would have thought that the animal was in distress. The call was so high-pitched, so plaintive, so primal, that it was difficult not to assign to it feelings of desperation. We have come to learn that male elk bugle for a number of reasons: to communicate their location, to challenge other males and establish dominance, and to attract females. They don't bugle from sadness or pain, as their wheezing,

warbling cry might suggest. I had to remind myself of this—
yet another reality of nature that turned out to be different
from what I initially suspected, yet another case of needing to
realign my thinking to reflect new information.

I quickly began looking around, as though I would be able
to see him—an absurd instinct given the distance the elk's song
can carry. I shook my head, then sat down for a couple of minutes on a nearby log. As the wind chilled the sweat on my back,
I heard the elk bugle once more, the sound washing over me as
he delivered a message to another member of his herd. I closed
my eyes and strained to detect all of its varied tones, knowing
full well that this vocalization contained pitches my ears may
never be able to hear.

OCTOPUS WORSHIP

I walked into the Heraklion Archaeological Museum knowing only that its website proclaimed it "the museum of Minoan culture, par excellence." I was a week into my month-long artist's residency in Crete, and a deep dive into the complex civilization that occupied this Greek island between 1900 and 1400 BCE seemed warranted. I'd been to a few small museums already and had seen a handful of Minoan snake goddess figurines, clay bull horns, shell jewelry, and, of course, lots of amphorae—classic ceramic vessels with narrow mouths and two handles. I expected to see plenty more of these artifacts; what I didn't expect was that seven of the amphorae in the collection would launch me into serious metaphysical contemplations.

I had been peering into the slick glass display cases, reading the bilingual interpretive signs, and dodging groups of tourists for about half an hour when I spotted the first vessel. From where I stood, I assumed I was just looking at a standard cream-colored jug adorned with a wavy line motif. But the wavy line had a bunch of little hollow circles running along one side of it. *It looks like an octopus tentacle*, I thought to myself, surprised at the resemblance. I pushed my way through the crowd to get a better view. It didn't just look like an octopus tentacle; it *was* an octopus tentacle. And there were eight of them, radiating from a body punctuated by two big

cartoon-like eyes. The design was so stylishly curved and finely edged that, in another setting, I might have mistaken it for a microbrewery logo or a skateboard deck. But this graphic wasn't created with Adobe software; these undulating arms and sensuously rounded bodies were painted by a Minoan around thirty-five hundred years ago.

After I spotted the first octopus amphora, I stepped a little to my left to look through the glass from a different angle. There was another. The jar itself was shorter and squatter, but the imagery was similar: a boldly painted octopus with swirling, sucker-clad appendages and googly eyes. I walked over to the next case and found two more octopus representations, one of which showed the creature lying on its side. By the time I got all the way through the museum, I had found seven in total.

I thought seven renditions of a somewhat rarely witnessed sea creature was a lot. As I stood in the spotless Minoan Pre-Palatial period room, I had to wonder if, during this chunk of history, the octopus was having a moment.

The octopus is definitely having a moment here in the Anthropocene. Following the popularity of the 2020 Netflix documentary *My Octopus Teacher* and Sy Montgomery's 2016 book that explores her encounters with captive cephalopods, *The Soul of an Octopus*, news about this fascinating creature's distributed neural networks and intriguing intelligence has spread far and wide. Many people know a handful of fun tidbits about octopuses these days—like the fact that they can change colors almost instantly to blend in with their surroundings, or the fact that their large, surprisingly vertebrate-like eyes operate independently and dilate on demand.

A month before I flew to Crete, I'd visited Monterey, California's, famous aquarium where I had to elbow my way through eight or nine people to secure a spot in front of the giant Pacific octopus tank. When I got there, I had an intimate view of the creature's tube feet—or suckers, as many people like to call them. They were quarter-sized white discs, grouped in pairs, running the length of each arm. He pressed them, two by two, inch by inch, into the glass. After contact, they expanded, looking a lot like the suction cups I once used to adhere my fifth-grade stained-glass projects to the kitchen window. Only there were dozens of them, moving in synchrony. We all watched him traverse his space by shooting one leg way out in front of his body, sticking his tube feet to the tank wall, and then, in one graceful but vigorous motion, hurling the rest of his body forward to catch up with his tentacle. "Whoa," the ten-year-old in front of me said as a mass of protoplasm surged across the tank. "He's, like, a superhero." Then, as if to prove that, the octopus clasped a white rock and promptly shimmied his patchy brown skin into a pale beige pattern that made him significantly more difficult to see. Everyone gasped.

What is it about the octopus draws us in?

The first time I saw an octopus in the wild, I was a seventeen-year-old snorkeling in St. Croix's tropical waters with my younger brother. We were both certified to scuba dive by then, and on our family's multiple spring break trips to the Caribbean over the years, we'd been fortunate to see an incredible diversity of underwater life, from eels and stingrays to parrotfish and puffers. But that afternoon, we were just cruising around with masks and fins in front of the hotel when I spotted a dark blob undulating in a crevice below me. If I hadn't known what to look for, I might have dismissed it as an anemone or a concentration of sea cucumbers. Its motion caught my attention, however. Anemones and sea cucumbers don't move. Fish do, but something about this creature's slither was distinctly unfishlike. Fish have a very metered method of swimming, like a predictable pop tune. What I was seeing was a syncopated collection of wiggles, crawls, and pauses—something much more like a free-form and flowy jazz composition. The only creature I could think of that navigated the world with such slinky style was the octopus, but according to my father, no one ever saw them during the day. I took a couple of kicks to get closer, looking for a telltale identifying feature. Sure enough, I spotted one: its eerie eyes—organs that, I would later learn, are comparable to ours.

I would like to say that our gazes locked, but I don't know enough about the focusing mechanisms of the octopus eye to do so. I can say that I was transfixed. For a second, I saw in another life form both radical difference and awe-inspiring similarity.

I pushed up to the surface and yelled for my brother. "Dude, dude! Get over here!" He took a stroke toward me. "Octopus! There's an octopus right down below me!" Behind the thick glass of his diving mask, my brother's eyes seemed to double in size. He put his head into the water and immediately started swimming away. He was only twelve years old at the time, but I've never let him forget that he fled.

Age aside, his reaction reminds me that creatures who are simultaneously very like and unlike us can also inspire fear.

I suspect part of this all-too-common human fear of the octopus is rooted in the spookiness of its eyes. For starters, they're quite large. Octopus eyes range from seventeen to twenty millimeters in diameter, while human eyes are about twenty-four millimeters wide. When you consider the differences in our body sizes and weights, the octopus eye is, relatively speaking, huge. Perhaps more significantly, though, the octopus eye has a pupil that is often described as "dumbbell shaped" or "rectangular." That pupil can look a bit creepy. However, it allows octopuses to see color, even though their eyes contain only one type of cone (ours have three: one for red, one for blue, and one for green) and therefore were assumed not to perceive color for many years.

Recent research has revealed that the octopus's unusually shaped pupil creates "chromatic aberration"—that is to say, it spreads out the wavelengths of light, functioning somewhat like a prism to split white light into its colored components. Because octopuses can change the depth of their eyeball (a skill that would have kept me from forty years of contact lens expenditures), they can choose to focus on one particular wavelength. Doing so allows them to "see" color through a mechanism that's nothing at all like the human rods-and-cones one. In addition, the octopus pupil can go from fully dilated to pinpoint in under a second—a great adaptation for quick diving and for emerging from a dark den into brightly lit shallow water.

Had I been able to watch the St. Croix octopus for long enough, I would have noticed that its pupils stayed horizontal regardless of its body position, thanks to organs called stato-cysts that, like our inner ear, measure and maintain equilib-rium. Their eyes operate independently of one another and can take in 180 degrees' worth of visual information, leading some people to describe them as "two panoramic cameras."

Comparing an octopus eye to a camera is totally legitimate; like a camera, it takes in light through an opening, uses a lens to focus that light, then projects that light onto photosensitive receptors. Vertebrate eyes, including our own, are also cam-era eyes. These days, it's generally accepted that the similarity between octopus eyes and vertebrate eyes is an example of convergent evolution—meaning that each species' eye devel-oped independently over time. We're just too different and too far apart, evolutionarily speaking, to have had a common camera-eye ancestor.

And yet, here we are, with these remarkably similar sense organs, using them to gaze out and investigate our separate and conjoined worlds—two sets of eyes that are similar enough to see themselves in each other but different enough to seem just a little bit alien.

Despite my own myopic-but-functional camera eyes and keen human-style depth perception, I've had a hard time depicting an octopus on a three-dimensional surface.

Every year, I make a woodblock carving of an inspirational creature to print and send as a New Year's card. For a long time, I gravitated toward plants or animals whose images were easy to reproduce—sand dollars and sea stars, for example. But a couple of years ago, my fascination with the octopus led to an attempt at portraying one. After pulling a photo off the internet and sitting down to work, I found that just sketching the creature's outline was tricky, let alone planning and executing cuts that might signal the possible contortions of eight independent legs. I started by drawing the octopus's mantle—the part we humans are tempted to call a head, although it contains digestive organs in addition to eyes—knowing full well that it changes shape constantly. I made lots of little circles for the ubiquitous tube feet, then decided that vaguely parallel swooping lines worked for the legs. I added a random hatch pattern on the mantle as a gesture toward the mottled coloring octopuses can take on, and I attempted to render that strange oblong pupil in its oversized eye. The design came out well enough to use, but it looked flat to me. It didn't really capture the creature's liveliness. I wasn't as proud of that card as I had been of previous ones. Of course, sand dollars and sea stars are far more fixed in their structure and live far less mobile lifestyles. They're also symmetrical.

Technically, the octopus is symmetrical as well, but you'll never find one assuming a left-mirrors-right position outside of a lab or an aquarium mortuary. Even then, you would have to manipulate and tack down a lot of slimy flesh to get that visual. Anyone who has seen cephalopods in the water knows that they are constantly shifting positions, reaching, and retracting. They never pose in any kind of a fixed and orderly fashion. As creatures of movement and exploration, they defy our desire to pin them into place.

Of the seven amphorae I saw in the Heraklion Archaeological Museum, the vessel I was most drawn to depicted the octopus in an arrangement that was, for the most part, bilaterally symmetrical. I say "for the most part" because, curiously, it had only seven legs. There were three on the left, three on the right, and one that came down from the center of its body. If that one leg had split in two, order would have reigned. Instead, it veered to the looker's left, throwing the arrangement slightly off balance. I didn't notice this detail at the museum; in fact, I didn't notice it until I downloaded my photos from my visit in preparation for making another carving in the Cretan studio I'd been granted access to through my artist's residency. I didn't quite know what I liked about this particular amphora when I snapped the picture; I just knew it was my favorite. But once I noticed the tentacle arrangement, I had to wonder: Why seven legs?

It's possible the artist just made a mistake, but given the obvious amount of time and care that went into the amphora's creation, that seemed unlikely. It's also possible that seven was a significant number to the Minoans. But after some thought, I decided that this "imperfection" was part of a plan.

I want to believe that imperfection is beautiful, since I embody it. I'm betting this Minoan artist felt the same way and perhaps painted a seven-legged octopus to give it a "blemish."

Or maybe the artist knew that octopuses can lose and grow back a limb. Maybe the artist, like me, found that capacity to be both fascinating and enviable—not to mention totally foreign to us. Or the artists may have believed that showing a marine creature mid-regeneration would be inspiring, since we humans are so often in the process of growing and changing. This would have enabled the painter to create some common ground with an otherwise alien-looking animal—the same thing I tried to do when I made my block prints.

While they don't have spinal cords, octopuses do have something like nine brains. This is another fun fact that is frequently repeated—no doubt because the concept is so difficult for us to wrap our singular brain around. This detail is often paired with a sentence like, "Two-thirds of an octopus's five hundred million neurons are located in their arms and body." Because of this unusual nerve cell distribution, each of the octopus's eight arms can actually move, grasp, taste, feel, twist, and smell independently. In fact, arms that have been severed from the rest of the body can continue to perform these tasks for at least a little while after they've been detached. (Even I might swim madly in the other direction if I saw a detached octopus limb inching my way.) Some people who work closely with octopuses claim that their different arms have different personalities. That idea, along with the previously held belief that there was little or no coordination among the tentacles, led to the broadly popularized "something like nine brains" theory.

Recently, however, results from maze-based experiments have suggested that octopus limbs may be less independent—"more networked," as scientist say—than previously thought. Tamar Gutnick's Okinawa Institute team taught one octopus arm to associate a particular maze door with a food reward and another door with no reward. When a different arm was offered the same two options, it repeatedly chose the door with the tasty treat, even though the octopus could not see the first arm in the process of making its selection. Somehow, the knowledge moved from one arm to the central nervous system and back out to the other arm. The arm-brains were connected.

It's looking like octopuses possess a complex combination of centralized and peripheral intelligence. We humans have absolutely no idea what that might feel like. As a result, while we find octopuses both curious and strange, we continue to evaluate them using the only kind of problem-solving games we know: the ones we use to measure our own mental aptitude.

And it turns out that octopuses can solve a lot of the problems human brains offer them. In addition to learning their way through mazes, octopuses have been witnessed crawling into and out of a variety of strangely shaped containers (broken bottles, for instance), unscrewing jars, poaching crabs from fishing pots, and tricking their prey. They have been known to "play" with non-food-related items in their tanks, seemingly just to find out more about the foreign objects. They have also been filmed collecting shells and other items from the seafloor to create walls around their dens—evidence of the kind of tool use we usually associate with vertebrates.

Octopuses can also learn to recognize faces, including human ones. In multiple laboratory settings, they have been able to distinguish staff members who reward them with food from identically dressed staff members who do not. Like most humans, they perk up when people with treats approach.

Yet these signs of intelligence don't exist because we have a common ancestor. Octopus and human neural networks are much too different to have a shared evolutionary path. Intelligence of this magnitude, like camera eyes, evolved twice, along two distinct paths. In the words of invertebrate scholar Peter Godfrey-Smith, author of *Other Minds: The Octopus, the Sea, and the Deep Origins of Consciousness*, an octopus is "probably the closest we will come to meeting an intelligent alien."

Is that why we can't seem to tear ourselves away from the aquarium's giant Pacific octopus tank?

It seems fair to assume that the Minoan artist who painted my favorite amphora had no knowledge of the many octopus facts we've come to accept as truth. As far as we know, the Minoans had no concept of a neural network, no understanding of the electrical impulses behind thought and movement, and no awareness of the molecular building blocks of life. They certainly weren't able to explain the biochemistry of the octopus's color change or the convergent evolution of human and octopus eyes. Yet these amphorae suggest that the Minoans put the octopus in a position of respect.

While I didn't exactly inventory the animals depicted in the Heraklion Archaeological Museum's galleries, I did notice which ones appeared where and how often. I expected lots of bulls, of course, given that they were the symbols of the Minoan royal lineage and culture. Other terrestrial creatures included a variety of birds, a bunch of beetles, and a fair number of goats. In the ocean-dwelling department, there was the famous dolphin fresco from the palace of Knossos, which has been largely reconstructed (and somewhat creatively imagined) from a few plaster fragments; a few fish (not as many as I would have expected, given that fish was and is a Cretan dietary staple); and a handful of a tritons (a cone-shaped mollusk with a spiral pattern on its shell). No sharks, no sea fans, no rays—but a cabal of octopuses. This distribution could be random, but I doubt it. Something about the octopus caught the Minoan eye and held it, as it had mine.

If I put myself into the Minoan equivalent of a woman's bathing suit and wade into shallow Cretan waters, what do I see?

I imagine I'm spotting a shape lurking in shallow water. I probably need to lance it with a spear and yank it out of the sea to get a good look at its sucker feet and giant slit-shaped pupils, since dive masks haven't been invented yet. As I run my hands along its legs and stick my fingers inside a couple of its myriad circular protuberances, I must be able to see how these features play a part in the creature's stealthy gliding motion. After watching the mysterious animal for a bit, I experience its transformation from a slow, side-moving, exploratory bottom crawler to a streamlined, super-fast swimmer intent on its destination. The Minoan me might think that this animal is both like a fish in its linear directionality and like a snake in it multiplanar side slithering.

Maybe I would conclude that the octopus's unique way of moving through the marine world, combined with its adaptability, speed, and one-of-a-kind presentation, makes it special—special enough to be painted onto an amphora. Perhaps even special enough to be treated like a god.

The two types of octopus movement that the Minoan me would have noticed are generally referred to as "swimming" and "crawling."

When an octopus needs to move rapidly—to escape from a predator, for example—it swims. Or, really, it propels itself in an action that, to us, resembles swimming. Because its mantle has a cavity inside it, water can be sucked into this space and then forcibly ejected through a siphon, thrusting the octopus in the opposite direction.

Swimming is very energy intensive, so crawling is the octopus's preferred method of locomotion. And thanks to all of the neurons in its suckers, an octopus is exploring its world when it's crawling. It's not "just" moving; it's having a sensory experience. Its suckers are one hundred times more sensitive to chemicals dissolved in water than our tongue is. It's thought that octopuses can taste pain in their prey, changes in the drug doses of their human caretakers, and stress in any living creature they encounter.

I can only exist in the ocean for an hour or so at a time—and that's with forty pounds of expensive, highly engineered equipment strapped to my body. If I could taste the seawater flowing around me, would it read like a map, a spy novel, or a guidebook? Would it be like hearing in an impressionist painting or smelling a symphony? Would I have an informational experience or a sensory one—or would those experiences be one and the same?

It seems that they are for the octopus, who, according to Peter Godfrey-Smith, "lives outside the usual brain/body divide." Where is that, and what would it be like to dwell in that space?

I don't think I get to know the answer to that question. Knowledge like this, to me, exists in the realm of the divine. Come to think of it, "divine" is the term we commonly apply to beings and forces that live outside the brain/body divide, entities we consider transcendent.

As I sat in the studio being serenaded by a herd of Crete's ubiquitous bleating goats, I moved my pencil over my woodblock, trying to replicate the shape of the creature I'd started to refer to as a "heptopus" to respect its seven-legged status. My favorite of the museum's amphora octopuses has an unquestionably female sensuality about it—human female sensuality, that is. It's curvy, with a body like a pomegranate and a head like a set of wide hips punctuated by big eyes. Between the two orbs of flesh lies a narrow waist—the cephalopod equivalent of the hourglass figure. The top two tentacles run alongside the octopus's body, mimicking and accentuating that figure. The next two down splay out to the sides and end in ornate spirals, flourishes like rings on fingers or polish on nails. The remaining three create something like a base for this bottom-heavy form. Much like the terra-cotta fertility goddesses I saw throughout Crete, this octopus is, in short, voluptuous.

Can that word can be applied to an invertebrate?

The more I sketched and contemplated my subject, the more convinced I became that the Minoan artist was actually making a goddess. Clearly, I was, too. And I was giving her human characteristics—trying, in my own way, to make her less alien. I suppose we humans have a long tradition of creating divinities who balance alienness with familiarity. It seems we need our inspirational figures to be similar enough that we can see ourselves reflected in their glow, yet different enough to serve as models we can stretch toward.

I smiled and started carving, pleased to be participating in this long lineage of deification.

There's yet another quote by Peter Godfrey-Smith that's been making the rounds lately. "An octopus is an intelligence without a self," he says in *Other Minds*.

I mentioned this to a friend who thought it sounded insulting to octopuses. "What, do octopuses not merit a self?" she asked. I, on the other hand, thought it was complimentary. To me, the idea of possessing intelligence without having to deal with the grasping, narrative-making, and very slippery construct we call "the self" sounds fantastic. I imagine being able to observe, collect and process data, understand, react, problem-solve, create—and even wonder and appreciate, if pure, selfless intelligence can do that—without the baggage of the "I" that all too often sits at the center of my universe. I imagine being more focused on my senses, my surroundings, and my experiences and less focused on what everything means to me and how it positions me among other members of my species.

What a relief that would be.

Octopuses live in every one of our oceans, occupying frigid waters in the Antarctic as well as tropical waters near the equator. A couple of specialized adaptations enable this.

One is their blue blood, the result of a copper-rich chemical called hemocyanin that, like our hemoglobin, bonds to oxygen and transports it around the body. Hemocyanin tends to hold more tightly to oxygen when temperatures drop, so Antarctic octopuses typically produce more of it.

Octopuses also adapt to colder temperatures by using RNA editing to change the way their neurons work. For years, scientists had assumed that Antarctic octopuses had evolved genes that allowed their nervous systems to function effectively in cold water. However, when researchers compared their DNA to the DNA of tropical octopuses, they found no differences between the molecules. So they looked to the creatures' RNA—the "translated" version of DNA that serves as a blueprint for the creation of proteins in all living things. Much to their surprise, they found that the cells of cold-water octopuses edit their RNA on the fly to directly affect how the building blocks of their nervous systems operate in frigid temperatures.

Octopuses don't have to change their DNA to contend with the cold; they just change the way their proteins are made or produce more hemocyanin. In other words, they don't have to evolve over generations and thousands of years to adapt to new conditions. They don't have to migrate thousands of miles to new habitats. And they don't have to change their environments. Instead, they change the way their bodies work. They change themselves.

As we humans face what may be our biggest collective challenge in our brief three-hundred-thousand-year history—our global climate crisis—I wonder if this flexibility might be just the inspiration we need to shift our thoughts and actions in solution-oriented ways.

In *The Soul of an Octopus*, Sy Montgomery discusses the theory that octopus intelligence evolved for different purposes than human intelligence. As she presents it, we needed—and need—our brains for socialization. To survive, human beings have had to work together, and that means communicating verbally and nonverbally. It also means we need to have some concept of what other individuals of our species are thinking and feeling. Scientists call this a "theory of mind." Octopuses have a theory of mind as well. However, Montgomery explains, theirs developed as a response to the amazing complexity of the environment they live in. They are both predator and prey at all times, and they constantly coexist with hundreds of other species. As Montgomery reminds us, we humans are lucky if we interact with one or two nonhuman creatures in a twenty-four-hour period. And the interactions we have with members of our own species aren't always going so well right now—especially when some of their beliefs differ from ours.

If we could develop a theory of mind that is less focused on the self, could we change ourselves—like the octopuses do—in response to our climate crisis? We clearly cannot wait for our species to evolve over generations before we put a stop to our planet-warming behaviors. Perhaps we could edit our narratives of what it means to be human, translating our DNA into an outlook that prioritizes sensory exploration and the health of all living things.

Along the way, I imagine that the ability to think in the octopus's more dispersed, decentralized fashion would precipitate a few other behavioral changes. It might make it a lot harder to eat octopus for dinner, tear up the reefs they call home, or dump toxins into the ocean they live and breathe in, for starters.

When I imagine myself into the past, I feel most comfortable wandering in the mindsets of pagan cultures. Archaic societies that worshipped the sun, deified rain, prayed to clouds, and held ceremonies on mountaintops all make sense to me. In the Heraklion Archaeology Museum, there are numerous mentions of "peak sanctuaries"—high points on the island of Crete where ancient Minoan artifacts have been unearthed. I drove to a few of these high points, and I was transported by their sweeping views of the rugged landscape, the endless sky, and the extensive sea. It makes sense that these were ceremonial sites in ancient times. They still are for me, even if many other humans worship down below in one of Crete's three thousand or so dark, incense-filled stone and concrete churches.

Pagan cultures also tended to celebrate—and even deify—the animals they interacted with. Their art typically includes depictions of the creatures they relied on for food. But other species—undomesticated ones, rare ones, the kinds that we still stop and take notice of when we see them in wild—also show up.

Case in point: the octopus. It has made appearances in belief systems from the Pacific Northwest to Japan, but it features most prominently in the spiritual traditions of Oceania. Kanaloa, the Hawaiian god of the underworld and the subconscious, often assumes the form of an octopus. The island of Moʻorea, near Tahiti, is home to a well-known octopus church. An octopus features prominently in Maori mythology, and the religious tradition of the Gilbert Islands includes an octopus god named Na Kika.

A belief system with an octopus at its center would work for me. It would honor the power of the senses and the wonder they enable us to take in. It would praise soft curves and eschew hard edges, preferring mountains, lakes, and ocean waves over rectilinear buildings and right-angled altars. It would prioritize color gradations, patterns, and the constant change of camouflage over clearly delineated hues. It would celebrate movement, dynamism, and flexibility. It would honor an intelligence that's anchored in the body and that holds the self loosely. And it would celebrate a living thing who changes itself in the face of shifting environmental conditions instead of continuing to change everything around it.

I wonder if this is why the octopus is having a moment—if there are other humans who, like me, are using their sensitive and intelligent tentacles to feel around for a new worldview. Perhaps they, like me, are finding one that is alien enough to spark radical change but familiar enough to feel accessible.

Which begs the question: Are we ready to model ourselves after a marine invertebrate? This one has been roaming our shared planet for 330 million years. Something's working.

As I stood in the museum looking at the octopus amphora, I remembered that "octo" is a Greek prefix. Indeed, "octopus" means "eight-foot" in Greek. I smiled at the convergence: There I was, in Greece, looking at an ancient Greek depiction of a creature the English language has given a Greek name. Before I left, I held up my hand and imagined the Minoan me in the Knossos pottery studio drawing a brush across a curve, matching (mostly) the left side of my octopus to the right, bringing to life my own octopus divinity, and making a wish for myself and my fellow human beings from all places and times.

Then I went back to my hotel, changed into my bathing suit, and ran down to Heraklion's waterfront. I didn't think I would see an octopus while doing laps along the pier, but since there were four or five men perched on the rocks fishing, I figured I would see something in the water—something different, something beautiful, something inspiring.

I stashed my shorts, shoes, and towel behind a concrete piling and stuffed my hair into a bathing cap. Then I pulled on my goggles—the little eye cups that enable me to see like a human in the nonhuman watery world. Although I would be looking with decidedly human eyes, I vowed to see with nonhuman ones—or at least to try.

I stumbled over the slimy rocks until I was in waist-deep water. Then I took a full breath, plunged beneath the surface, and let the alien environment in—into my eyes, into my ears, into my skin. I hoped as I did so it was making me more a part of it, softening the edges of this self of mine that can feel so separated from the seas it navigates. Slowly, I started moving my four limbs, propelling myself through the water in the way that my human body knows how. I pointed myself away from the pier and swam into the current.

PINK HEALING

A couple of summers ago, I accompanied my mother on a bus tour of Switzerland. I don't love the bus-tour format for a number of reasons—too much windshield time, too little time to explore, too many other people with whom I struggle to connect, and too much pressure to interact with them. It's not an environment in which I can turn to the person in the next seat and tell them that I am crushed by the knowledge that we are losing 150–200 species to extinction every day and that we are implicated in this loss by riding around in a fossil-fuel-burning bus. But I enjoy traveling with my mother. Now that she's in her eighties, she's found that bus tours are the best way for her to keep seeing the world—a passion we share. So, despite the nagging climate guilt I experience every time I get on a plane, I flew across the Atlantic and joined her.

About six days into the tour, we encountered traffic heavy enough to slow our ascent of Julier Pass, a seventy-five-hundred-foot mountain saddle outside of the resort town of St. Moritz, to a crawl. I stared out the tinted window at the road bikers beating us uphill, squirming in my seat as I thought about missing the chance to plant my sneakers on another section of the Alps. I had been running every day since I arrived in the country, and this trail-based exploration had both introduced me to a new ecosystem and provided me with my daily fix of the wild world. While I gawked at the alpine

meadows flanking the highway and worried about my personal exercise schedule, the tour guests across the aisle were glued to their phone screens, seemingly oblivious to the stunning scenery passing us by. At the same time, the people in front of me were swapping tales about their worst birthday presents. I put my headphones on and pressed my nose to the glass, trying to absorb and contain everyone's allotment of natural beauty on my own—the serrated peaks protruding up from the ridgeline, the lush and expansive fields that lay below them, the cloudless blue sky that hovered over all of it.

I pulled off my headphones when the guide started explaining that Julier Pass had seen regular use since the Roman era. She said we would stop there for about fifteen minutes so that everyone would have a chance to get coffee. Despite our majestic surroundings, I felt the darkness—the nagging guilt that always lies dormant in my abdomen—spread from its hiding place into my limbs and lungs. We were about to idle for twenty minutes in a fifty-six-passenger bus that probably got a mile or two to the gallon, and there were five other buses in the pullout doing the same. Nearly everyone lingering outside of the café was holding a disposable coffee cup with a plastic lid, and the ones that weren't were clutching small single-use plastic bottles. The road, while it allowed us to access a spectacular location, was no doubt responsible for countless animal deaths—not to mention habitat fragmentation and toxic runoff from the petroleum products leaked by the myriad vehicles that traveled it. When we got to the hotel, there would be a lavish dinner that resulted in a massive amount of food and packaging waste. That dinner might be outside, under propane-powered lamps that heated the great outdoors. And to top it all off, I had traveled there on an eleven-hour flight. Assuming my airplane had produced carbon dioxide at the commonly cited rate of ninety kilograms per passenger per hour, I calculated a one-thousand-kilogram personal responsibility for the worsening of the climate crisis as a result of that one leg of my trip. I didn't even want to think about the amount of fossil fuels necessary to mine, assemble, and transport the materials that made up my ultra-fancy running shoes. I closed my eyes and forced myself to take a deep breath.

When I opened my eyes, I noticed patches of pink through the window glass. At first, they were small and nestled into the wrinkles between ridges. But as we approached the pullout at the top of Julier Pass, the fuchsia flower congregations grew larger. One of them almost completely blanketed the east side of the road. I squinted as the driver parked the bus and the individual plants came into focus. Each one was about three feet tall, with a strong central stalk rimmed by spear-like, dark green leaves. At the top of those stalks were flowers—anywhere between three and fifteen of them—in a columnar formation, looking a bit like hot-pink violets encircling a woody stem. I smiled.

"What is this stuff?" one of the birthday present people asked.

"I have no idea," another answered.

"Maybe carnations?" a third chimed in.

"Fireweed," I said softly. There it was, one of my wandering life's steady companions, at the top of a Swiss mountain, greeting me with its electric magenta embrace. I instantly felt at home.

"You know what it is, Bridge?" my mother asked.

"It's fireweed." As I said that, I heard several people ask the guide the same question. Though she knew an impressive amount about the construction of the Julier Pass, she didn't know what the florescence next to it was.

"Bridget says it's fireweed," my mother shouted over the din. I could have elaborated on that, explaining why it was called fireweed and how I knew about it. Instead, I shrank down into my seat, feeling like the kid who answered one too many questions in her fourth-grade class.

As soon as the bus came to a stop and everyone got out to line up for coffee, I grabbed my camera, looked both ways, and crossed the highway. I hopped from the edge of the asphalt onto one rock, then another, then across a gravelly ditch and into the sea of color. "Hello, my friends," I said, stooping down a bit to meet the flower's bracts at my eye level. "You're peaking right now, aren't you?"

I walked along the edge of the field, careful not to trample the stalks but eager to take pictures of the European version of a familiar alpine wildflower. I spent my whole fifteen-minute break amid the fireweed, inhaling the oxygen it was emitting, soaking up its color, and appreciating its presence. I felt my entire demeanor shift. The gloom that had threatened to overtake me on the bus had retreated. I was so lucky—lucky to be in a beautiful place, yes—but luckier still that I was connected to a plant in a country I had never been to before. This plant had the power to nourish me, to help me feel whole.

My history with fireweed began when I first started spending time in the mountains of Wyoming, back in the late 1990s when I was leading young adults on thirty-day wilderness hiking courses. There, fireweed is a summer staple at elevations of six thousand to eight thousand feet, and the combination of its color and its propensity for growing in dense patches makes it one of the ecosystem's most obvious warm-weather residents. It often grows on the sides of roads, so in addition to seeing it while I was hiking, it made regular appearances on my drives through the mountains. As my four-cylinder Mazda pickup truck chugged up a pass, I could raise my eyes from the ribbon of asphalt and let them rest for a few moments in a cheery field of flowers. I didn't know why fireweed always brightened my day; I suppose at the time I just chalked it up to its vibrant hue. Many years later, I learned that its gifts to the landscape went far beyond its visual splendor.

On a July day in 2014, I decided to ride the classic Cache Creek to West Game Creek mountain bike route outside of Jackson Hole, Wyoming. For the decade I'd lived in nearby Victor, Idaho, this had been one of my go-to rides, but I'd avoided it altogether since the Little Horse Thief Fire ravaged the area in September 2012. The event was still fresh in everyone's minds, having occurred back before forest fires in the northern Rockies became somewhat commonplace. I remembered the initial shock of seeing local smoke that summer—not just the kind that blew in every year from Oregon and Washington. And I remembered the firefighter camps on Teton Pass, where hundreds of tents and trucks filled the parking lots of my favorite trailheads. But two years had passed since then, so I figured it was safe enough—for both me and the land—to go back in there.

Right out of the parking lot, I was surrounded by thick foliage. Pedaling on the beginning section of the ride was like biking through an alpine jungle, complete with shoulder-high cow parsnip plants and giant horseflies. The vegetation was so thick that I could almost feel the humidity of its collective respiration on my own sweaty skin. Stems of all kinds bent into the trail, slapping my bare shins as I rode by. I slowly worked my way up from out of the creek bed and onto the switchbacks of the Ferrin's Trail, all the while moving through sections of the forest that the fire never reached. After about forty-five minutes, I arrived at the flat area that separates the Cache Creek drainage from the West Game Creek drainage, the spot from which the roller-coaster-like descent begins. I could see some burned trees out at the edge of my view, but otherwise, the place looked like I remembered. I stretched my legs, ate a few handfuls of pretzels, and pointed my bike down the West Game Trail, ready to bank its swoopy turns with a huge smile on my face.

After I crossed under the undamaged portion of the ridge, I rounded a corner and entered the burned zone. It felt like another world, and in a sense, it was. I had expected to see jet-black tree trunks with no foliage and no branches—the charcoal skeletons of former lodgepole and ponderosa pines—and I did. They were everywhere, standing like crispy matchstick sentinels, reminders of the forest's recent past. What I didn't expect were the giant spatters of fuchsia just beneath them. While I knew each patch of fireweed comprised hundreds of individual plants, their stalks were, at first, indistinguishable. But when the trail wove through one particularly large grouping, I jammed on my brakes and stopped to look more carefully. Each stem was at least four feet tall, with its top quarter devoted entirely to its showy, four-petaled flowers.

Although each of the flowers was small—an inch or so in diameter—I knew that a single plant could stack up to fifty of them up its stalk over the course of a season. Not all of those flowers would be visible at once, since fireweed blooms from bottom to top. But at any given moment, ten or twelve flowers would be open for pollination business. It hit me then how useful this strategy was for the propagation of the species. Continual blooming allowed pollination to occur throughout the summer, ensuring that short-term weather or insect events didn't affect the process. In addition, opening only ten or twelve flowers at a time ensured that not too much energy was invested in the flowering process at once, leaving enough for the plant to keep flourishing. I shook my head, ashamed that I had never really put this together before, then went back to admiring the flowers for their aesthetic properties. The ones at the top that were just beginning to bud looked like magenta raindrops lying on their sides. Meanwhile, below the blooming flowers, their fading brethren had dropped everything but their central spikes, which still stuck out straight from their stems like little pink candy straws.

On another trail at another time, I would not have stopped to take such a close look at an individual patch of fireweed. It's one of the most common flowers in the Rocky Mountains, and I saw it nearly every day from midsummer until the first snowfall, year after year. But I had never seen it smothering the scarred ground of a recently burned forest. It had a much more powerful impact on me that day—and not just because the contrast of pink and black was so stark. The fireweed's complete dominance of the area gave me pause. To stand there was to intuit the plant's mission: It was healing the land.

When living things move into inhospitable areas and take root, we put them into that loose classification of organisms we call "pioneer species." These species manage to eke out a living in avalanche debris, sand, glacial moraine, highway shoulders, and denuded ground. No matter how fertile a forest's soil was prior to a fire, really hot blazes like the Little Horse Thief Fire kill the majority of microorganisms that once lived in it, turning it from what is called "organic soil" to what is called "mineral soil." Fires also frequently leave the ground less porous and more compact than it was, increasing water runoff and decreasing the amount of moisture that can get to a plant's roots. And to top it off, they eliminate any and all shade. While these conditions are bad news for most vegetation, fireweed thrives in full, direct sun and ashy, nutrient-poor soil. In this scenario, it can take over without a hint of competition.

One of fireweed's main takeover strategies is massive seed production. Each plant and its fifty some-odd flowers can produce upward of eighty thousand seeds. That seems like an impossibly high number until late August, when every single stalk is covered in what looks like silky white fuzz. When I've rubbed this cottony substance between my thumb and forefinger, enjoying its smoothness on my roughened skin, I've marveled at how something that seems to the human eye like the withering remnants of a dying flower is actually a perfectly engineered seed-delivery mechanism. Each little hair is like a parachute that facilitates the flight and soft landing of the tiny dot—the seed—affixed to its end. In certain fields during late summer, fireweed pollen on the wing can look almost like a snowstorm. When an entire forest has been destroyed, this dispersal system ensures that fireweed has a good chance of survival, if not total dominance. Case in point: In studies of the land surrounding Mount St. Helens one year after its infamous 1980 eruption, 81 percent of the seeds collected were fireweed seeds. Fast-forward to two years after the Little Horse Thief Fire, and the result is the monoculture sea of flaming pink I witnessed on the Game Creek Trail on that day in 2014.

For years, I gathered fireweed seeds in an effort to cultivate the plant at my house on the west side of the Tetons. Having taken lupine seed pods from dried flowers near the forest boundary and successfully sowed them in a backyard bed, I figured I could do the same with fireweed. I was all for cultivating aggressive noninvasive species on my land; anything that kept the dandelions and thistles out of my beds and provided some color in the process was more than welcome. So for the first few years that I lived in that house, I filled a Ziploc bag with fireweed fuzz every autumn and scattered it over the barren spots between the daisies and the lupines, imagining the brilliant patches of hot pink that would punctuate the mounds of white and purple the following summer. One year, I had two stalks. The other years were total failures. In retrospect, I wonder if my

yard was too easy a place to live, if a place with loose, nutrient-rich soil and ample water was just not challenging enough
for fireweed.

Meanwhile, in a postburn forest, once some of those
millions of seeds hit the depleted mineral soil they love, they
germinate quickly—typically within ten days. When the
plants are established, they send out rhizomes. These horizontal, underground, stem-like structures quickly propagate the
fireweed and have the beneficial side effect of holding the area's
soil together. New stalks spring up from these rhizomes (this
is known as "vegetative reproduction"—a much more rapid
and efficient process than sexual reproduction), and flowers on
those new stalks can bloom within a month of emergence. Cut
or torn rhizomes send up even more new stalks, like the weeds
your mother told you never to yank on because they would
multiply before your eyes. All told, these qualities combine to
make fireweed a regeneration superhero. They're also the reason
for its name.

Fireweed density seems to peak in an area a couple of
years after it moves in. The plant remains a dominant force
in the landscape for another five to ten years, but after that,
it becomes less common. This is partly a result of shadier
conditions. Fireweed thrives in bright sun, and open areas tend
to diminish as trees expand their canopies. Another factor in
the gradual decrease in fireweed density is simply the increase
in competition. As the soil heals and becomes more organic
and nutrient-rich again—in large part thanks to the fireweed
growing in it and holding it together—it becomes habitable by
other species.

So, in a manner of speaking, the fireweed moves on. It gets
the healing started and then gets out of the way. It cedes its
space to others so they can thrive, while it moves on to pioneer

other questionable habitats. The remediation team comes to the rescue, always. That's the way it's happened for millennia. That's the way it should keep happening.

When the darkness overcomes me, however, I wonder if this will continue to be the case. I wonder if, as we all continue our current behaviors, the organisms who have healed denuded land, the sinks that have sequestered carbon, and the natural cycles that have mitigated our impacts will simply be overwhelmed by the amount of repair there is to do. Of course, Earth has recovered from wild temperature fluctuations, asteroid impacts, and radical changes in its chemical composition in the past. In some way or another, the planet will "recover" from whatever we are doing to it now. It's just that we don't know what that recovery will look like—or who will be here to witness it. It's fair to assume that healing process will not be as beautiful or as rapid as fireweed's repair of a Teton mountainside.

I go through phases when I think that the impending sixth extinction is not that big a deal, that in the grand scheme of our universe's enormous size and infinite lifespan—as well as our insignificant role in it—what we've done here is just another blip in the unfolding of time and space. But I also go through phases of extreme sadness, when I cannot wrap my head or heart around the amount of fascinating, vibrant, ingenious, crazy-looking, and wonder-inducing species we are about to lose. I think about the ones that have passed into oblivion in the last decade alone: the Pyrenean ibex, the Western black rhino, the Yangtze river dolphin, the Moorean tree snail, the Rocky Mountain locust. There are hundreds more, including some we probably don't even know about. During these phases, I feel empty and barren, like my interior is a fire-scarred landscape where no positive thoughts can take root. This is a very dark, very lonely place.

People always ask me why I look to nonhuman creatures to make sense of my life. The simplest answer I can give is to say that I know them, and I admire them. Yes, I feel connections to fireweed, sea stars, kelp, and owls, but I'm pretty sure they don't feel any connections to me in return. My knowledge of them doesn't evolve out of verbal communication, the way it does with other human beings. My knowledge of them comes from observing them, from giving them space, and from letting them be who they are. It comes, too, from reading about them and working to acquire the wisdom other humans have collected in their efforts to better understand nonhuman worlds. That knowledge nourishes me. It has led me to consider these creatures my equals and my sources of inspiration.

When I look around, I see so much creativity in nature. I see problem-solving strategies that address every condition imaginable. I see species experimenting and evolving. I also see them cooperating and, like fireweed, facing destruction and healing wounds. Most of all I see humility, patience, and endurance, as well as community orientation and long-term vision—traits I find lacking in my own species, and traits I know are required to maintain the health of our amazing planet. I suppose that seeing such a wide variety of other creatures working hard to fill their roles in our planet's myriad intertwined ecosystems makes me feel less lonely.

My interactions with nonhumans heal my personal wounds every day. I believe they have the potential to heal global wounds too. This is why I try to understand them on their own terms, why I try to help others appreciate their value: to help them speak their needs in a world that is increasingly managed by humans.

This is why I will dash from the bus into the field of fireweed when given the chance.

Right after I returned from that trip to Switzerland, I was back in Jackson Hole and running on the Phillips Pass Trail, another local favorite that leaves from a busy roadside parking area and meanders up into the Bridger-Teton Wilderness. I chose it because, two days earlier, I had been mountain biking one ridge east of Phillips Pass and seen big pink brushstrokes across its alpine meadows. "Holy moly, that's fireweed, over there, isn't it?" I said to my companion as we took a snack break at the best viewpoint on the trail.

"I'd say so," he replied. "Looks like it's right there below Phillips Pass, huh?"

The next day, I cleared my schedule to make time for the two-and-a-half-hour run I would need to do to access those patches. From the Phillips parking area, I plodded along up through the spruce and fir forest for about half an hour, then kept up a steady jog as I passed the big groves of mature aspen. Once I got to the creek bed that I knew led to the meadow I'd spotted on our bike ride, I slowed to a walk for the final steep uphill section. I looked down and noticed that the ground was dry, gray, and full of cobbles. I was entering a zone of low-quality soil: Cue the fireweed.

I actually heard it before I saw it. It wasn't the fireweed making noise, however; it was the bees. As I picked up my pace again, the buzzing got louder. When I finally crested the horizon line below the meadow, I came into a field of flowers larger and denser than any I have ever seen. I stopped running and just stood there, feeling myself be swallowed by pink. The fireweed was over five feet tall and growing with such enthusiasm that it nearly obscured the trail. To move forward, I had to push stalks to my side like a jungle bushwhacker—but I had to do it cautiously, as there was a bee on almost every one of those stalks. The insects looked fat and heavy, and I was surprised they could fly at all. Between their pulsating hums and sluggish flights, I took them to be drunk on fireweed pollen. And they weren't the only creatures around. Within the span of a few paces, I managed to count eight frenzied hummingbirds. I knew there were snakes, voles, and ground squirrels below me; red-tailed hawks, ravens, and Clark's nutcrackers above me; and copious quantities of oxygen and pollen being produced all around me. This community was thriving.

It occurred to me that it was thriving because of the sheer number of plants that were packed in together, the sheer number of plants doing the same thing with the same objective. I would not have embarked upon a two-and-a-half-hour run to see one fireweed plant. I ventured up there to see thousands—maybe tens of thousands—of individuals working together. It takes that many stalks of fireweed to make a thriving community. It takes that many stalks of fireweed to heal a damaged burn site.

It will take that many humans to heal a damaged planet.

I suddenly regretted not telling the people on the tour bus two weeks earlier about the wonder of fireweed. Maybe they had never had the chance to stand in a field of flowers so thick that they couldn't see the ground. Perhaps they had never lost the sound of their own heartbeat in the hum of hungry bees or mixed their respirations with the air exchange of tens of thousands of plants. They'll never know what this is like if no one tells them. I had forgotten that this is something I can do, something I can contribute, something that can begin to offset the impact of my presence on the planet. If I had offered a few more details about fireweed and shared my enthusiasm and appreciation for it, I might have, in some small way, recruited them into a healing role. And I might have made a connection with them, too. I might have felt less alone. *Next time*, I thought to myself. *Next time, I'll remember what it means to be on the remediation team.*

After spending some time wandering up and down the trail and sticking my arms into the fireweed forest on either side of me, I stopped thinking altogether and just stood there, looking, listening, breathing. If someone had been riding the trail on the next ridge over—the one I'd been on just two days earlier—they would not have been able to distinguish five-foot-eight-inch me from the five-foot-eight-inch fireweed stalks I stood in.

We were all standing together, performing our version of healing.

ACKNOWLEDGMENTS

No creative endeavor is a solo project, and I am fortunate to have had the support of many people in the making of this book.

For starters, I have the world's best editing buddy. This book would not be what it is without the suggestions and support of Courtney Kersten, a fine author in her own right, who somehow seems to understand what I am trying say better than I do. She helped me sculpt many of these essays into their polished forms and kept me on track with her unflagging positivity.

I owe a debt of gratitude to Andy Couturier for getting me started on this path, encouraging me along the way, and including me in his vibrant writing community, theopening. org. Thanks also to Nicole Walker and the Northern Arizona University MFA cadre for their teachings and guidance. My writing buddies Miranda Perrone and Addy Santese provided helpful feedback on early versions of some of these essays, and editor Elizabeth Witte assisted me in shaping several others.

Thank you to all of my editing clients, past and present, who teach me so much about the art and craft of writing by sharing their work with me.

I greatly appreciate the support of the literary magazines that have published my work in the past; they are listed in the section of this book titled "Publication Credits." I must also thank the regular readers of my blog (exploraspective.

wordpress.com) for keeping me engaged in a project that, for ten years, has served as a sandbox in which I explore ideas that often evolve into more formal pieces of writing.

The US Fish and Wildlife Service provided generous support for the research and writing of "Negotiating Fluidity" through their Voices in the Wilderness Writing Residency. Julienne Dolphin Wilding and the Event Horizon Residency in Crete provided me with the space, time, and inspiration to write "Octopus Worship" and to edit several of the other essays in this book.

It is because of Marguerite Avery's commitment to my work that this book exists in its current form. Huge thanks to her and everyone at Texas A&M University Press for supporting environmental writing in general and this project in particular.

Ronna Leon and Arts Benicia generously provided me with the space, equipment, and expertise to print the linoleum block illustrations in this book.

I am eternally grateful to my parents, who, whether they have agreed with my choices or not, have nevertheless unfailingly supported my curiosity, love of adventure, and commitment to exploring the natural world. In doing so, they have empowered me to live the experiences that inspired these essays.

Finally, I must thank the myriad nonhuman creatures with whom I interact for their constant presence in my life as sources of wonder, wisdom, and emotional sanctuary. This book is a tribute to them. May our species begin to do what it takes to ensure their continued existence on our planet.

RESOURCES

TANGLED

The Monterey Bay Aquarium's website provides useful introductions to both giant kelp and bull kelp as well as all of the other species that inhabit the central coastal waters along with them. The site also hosts a "kelp cam" where you can get a sense of what kelp look like from underwater, when they are still part of a forest rather than washed up on the beach. The website asknature.com hosts the article "Highly Stretchable Stipe Resists Breaking," which has a great explanation of why kelp are so stretchy. And if you would like to connect with both the aesthetics and the science of kelp, Josie Iselin's book *The Curious World of Seaweed* is inspiring.

SOLITUDE IN DENSITY

One of the best resources on monarch butterflies in California is the website of the Xerces Society (xerces.org), a nonprofit organization dedicated to invertebrate conservation. An entire section of the site is devoted to its Western Monarch Conservation initiative, and it includes information about overwintering sites, milkweed habitat, butterfly-friendly gardens, annual Thanksgiving counts, and legal initiatives aimed at protecting monarchs. There are also links to numerous white papers and yearly population statistics. What you can't find on that page can almost certainly be found in the "Press and Media" section of the website.

BENEATH THE SURFACE

A thorough description of humpback whales' appearance, habits, and survival threats can be found on the National Oceanic and Atmospheric Administration (NOAA) Fisheries website in the "species directory." An article on NOAA's National Marine Sanctuaries website called "Unraveling Mysteries of Humpback Whale Song" offers an excellent summary of the state of humpback whale song research. I was alerted to the transmission of whale vocalizations from population to population by the *New York Times* article "Humpback Whales Pass Their Songs across Oceans," based on a paper published in *Royal Society Open Science*.

If you've never listened to humpback whale song, I highly recommend that you do so! You can find many recordings on YouTube; however, bio-acoustician Roger Payne's 1970 album *Songs of the Humpback Whale* is the most well-known and significant collection of whale vocalizations. When it was released, it sold more copies than any other nature recording in history and raised awareness about whale hunting—leading to the passage of the Marine Mammal Protection Act in 1972.

WHEN THE CHESTNUT FALLS FAR

The American Chestnut Foundation's website, tacf.org, is an exhaustive chestnut resource, containing information about everything from the tree's history to the science and technology being harnessed in the attempt to restore it to the forests of the eastern United States. In addition, both NPR ("The American Chestnut Was Wiped Out a Century Ago: Could It Make a Comeback?") and the *New York Times Magazine* ("Can Genetic Engineering Bring Back the American Chestnut?") have produced thorough and well-researched summaries of the chestnut's recent status.

OWLGAZING

A great basic resource on great horned owls—and, indeed, on any bird—is the Cornell Lab of Ornithology's website, allaboutbirds.org. *National Geographic* has published a number of articles about this most commonly spotted owl as well.

An excellent source of information about owls and other raptors is your local raptor center. Raptor centers exist throughout the United States and beyond, typically as both rehabilitation facilities and educational institutions, and it's well worth the time and effort to learn about—and potentially support—the center nearest you. I've had the most exposure to the Teton Raptor Center in Wilson, Wyoming, an inspiring nonprofit that treats over 150 birds per year, runs field studies, and maintains a small flock of resident raptors who work to educate people about their species. As of the publication of this book, K2, the owl mentioned in this essay, had recently been retired from "glove work" after nineteen years of service in the Tetons. She now impresses visitors to the Tracy Aviary in Salt Lake City.

IBEX TROPHIES

The story of the European ibex population's recovery is recounted in several articles, including "The Improbable Return of the Mountain King," published by the Organisation of Swiss Abroad (OSA); and "Alpine Ibex Saved by an Italian King" on the Shepherds of Wildlife Society's website. More scientific information is provided in "Recovery of Alpine Ibex from near Extinction: The Result of Effective Protection, Captive Breeding, and Reintroductions," from the February 1991 issue of *Applied Animal Behavior Science.*

I consulted the website of the International Council for Game & Wildlife Conservation (CIC) United Kingdom Trophy Evaluation Board for the discussion of ibex trophies. Two articles on Swissinfo.ch (a website managed by the Swiss

Broadcasting Corporation), "Trophy Hunters Pay Thousands to Kill Iconic Swiss Animal" and "Foreigners Banned from Ibex Trophy Hunting," discuss the ongoing controversy surrounding the hunting of ibex by foreigners. Of course, the definitive source on ibex hunting regulations in Valais is the canton's website (vs.ch), where, if you speak French, you can also download and read the highly informative "Fiche factuelle boquetin."

Videos of ibex scaling near-vertical dam walls can be found on the YouTube channels hosted by the BBC and *National Geographic*.

ON SANDERLINGS, THE COLLECTIVE, AND ME

Once again, the Cornell Lab of Ornithology's website, allaboutbirds.org, is a great source of basic information about sanderlings. The British shorebird conservation group Wader Quest also has a thorough sanderlings entry at waderquest.net. Specifics about sanderlings' lack of flock cohesion are detailed in the article "Space, Time and the Pattern of Individual Associations in a Group-Living Species: Sanderlings Have No Friends," by J. P. Myers, in *Behavioral Ecology and Sociobiology*.

Many videos of the Mevlevi engaged in the *sema* can be found on YouTube and are well worth watching. Because the ceremony is considered a UNESCO Intangible Cultural Heritage event, photos and more information about the dance are provided at unesco.org.

The Grateful Dead documentary referred to in the essay, *Long Strange Trip*, is available on streaming services.

ECHINODERM ENVY

A 2001 article by I. C. Wilkie titled "Autotomy as a Prelude to Regeneration in Echinoderms" and published in *Microscopy Research and Technique* provides detail on this interesting sea star superpower. However, the foundational work on autotomy is a 1935 article, "Autotomy and Regeneration in Hawaiian Starfishes," written by Charles Howard Edmondson and published by the Bishop Museum. The amazing powers of the *Linckia* genus are described in the book *Whole Body Regeneration: Methods and Protocols.* Simply typing "sea star arm regeneration" into the YouTube search engine will yield numerous time-lapse videos of this fascinating process in a variety of species.

Media outlets from the *New York Times* to *Smithsonian* have published excellent articles on sea star wasting disease (SSWD); however, more detailed information and progress reports—as well as directions for submitting observations of SSWD cases in the wild—can be found at the University of California Santa Cruz's Multi-Agency Rocky Intertidal Network (MARINe) SSWD page (also linked at seastarwasting.org). As of the publication of this book, the National Oceanic and Atmospheric Administration (NOAA) had recommended the sunflower sea star be listed as an endangered species. The NOAA Fisheries website is the best place to follow the progress of that effort.

SPONGES, THE ANTIHUMANS

While I was in Roatan, I came by a copy of *Caribbean Reef Life*, by Mickey Charteris, which is the Caribbean diver's bible. In addition to including twenty-two pages devoted to sponges, it's a terrific resource for identifying and learning a few facts about all of the other organisms mentioned in this essay as well as many, many more. Charteris's accompanying website, caribbeanreeflife.com, features a blogpost titled "Redwoods of the Deep: Giant Barrel Sponges May Live to Be over 2000 Years Old!" that includes a photo of the Texas barrel sponge, some great information about how to age barrel sponges, and a discussion of threats to the species. The post also contains an embedded version of the video depicting a sponge siphoning dyed water. The University of Michigan's Animal Diversity Web (ADW) is a great resource for the basics on just about any animal, but I found its entry on *Xestospongia muta* to be especially helpful.

NEGOTIATING FLUIDITY

This essay is based on my participation in the Voices of the Wilderness Artist's Residency program, a terrific endeavor that pairs artists with Alaskan public land management agencies in order to facilitate communication of these agencies' projects to broader audiences. I was placed with the US Fish and Wildlife Service (USFWS) in the Arctic National Wildlife Refuge, and, in exchange for their support for my field time, I wrote an extensive blogpost about my experience. This article, called "Chasing Eiders," includes some of my photos and can be found on the USFWS blog, hosted on the platform *Medium*. Much of the content in that post and in this essay came from my observations, voice recordings, and videos of our daily activities.

I consulted a host of scholarly articles about Arctic melting trends in order to write this essay; however, the pace of change in the far northern latitudes is so rapid right now that even recent papers are already outdated. For that reason, I recommend the National Snow and Ice Data Center's site, "Arctic Sea Ice News and Analysis," and the National Oceanic and Atmospheric Administration's (NOAA) website, climate.gov, for the most up-to-date information on Arctic warming trends and their terrifying effects.

A helpful discussion of Fata Morgana can be found on the SKYbrary website, skybrary.aero.

You can find out more about the *Infinity*'s expeditions and track their progress at infinityexpedition.org or on their

Facebook page. The TV series made from footage of their journey, called *Expedition to the Edge*, can be accessed via the Discovery Channel's website. For more information about the Earth Flag, see flagofplanetearth.com.

RIPPLING LINES

Most of the facts about stingrays included in this essay can be found on any website dedicated to ocean creatures. That said, when I'm in need of fish identification assistance or fun fish factoids during Mexican ocean-based adventures, I consult the website Mexican-fish.com. It combines helpful photographs with exhaustive description and specifications. The *National Geographic* article about Huntington Beach's increase in stingray wounds referenced in this essay is called "What's Behind a Surge in Stingray Attacks?" and was published in January 2018.

There is a plethora of information about Steve Irwin's death on the internet, including articles published in media outlets ranging from the *New York Times* and NPR to TMZ and *People*. YouTube hosts hundreds of video snippets of his animal encounters. Among the ones I watched were "Steve Irwin's Biggest Crocodile Battles," "Steve Irwin's Best Moments: Venomous Snakes," and "Steve Irwin Catches a Crocodile in Epic Battle," all of which can be found on the RealWild channel. However, any montage of his work over the years will give you a sense of both his enthusiasm and his unique way of interacting with animals. To learn more about the creation and administration of the Steve Irwin Wildlife Reserve, go to the Australian Zoo's website.

LOSING REFUGE

Journalist Todd Wilkinson has written extensively about a variety of issues facing the Greater Yellowstone Ecosystem, including the controversies surrounding the supplemental feeding program on the National Elk Refuge. His articles in *Mountain Journal* and the *Jackson Hole News and Guide* were particularly helpful in the writing of this essay. Steve Moriss's article "National Elk Refuge, 1912–2012" on the Jackson Hole Historical Society and Museum's website provided important context for the discussion. Both the Jackson Hole Wildlife Foundation's website's feeding program updates and the Wyoming State Department of Fish and Game's website's CWD and brucellosis status updates allowed for the construction of an event timeline. Earthjustice's website contains additional information about the lawsuits the organization filed against the US Fish and Wildlife Service, which is itself an important source of information, both through its website and in person through its Visitors' Center and seasonal refuge tours.

Finally, articles in *New York* and in the *New York Times Magazine* were helpful in filling in the gaps in my personal knowledge of the John Friend/Anusara Yoga scandal.

OCTOPUS WORSHIP

This essay mentions several recent octopus-centered works that inspired its writing: *The Soul of an Octopus*, by Sy Montgomery (Simon and Schuster, 2016); *Other Minds*, by Peter Godfrey-Smith (HarperCollins, 2016); and *My Octopus Teacher*, a 2020 Netflix documentary by Pippa Ehrlich and James Reed. I cannot recommend them highly enough. In addition, a number of scientific articles provided me with information critical to my understanding of these complex animals, including "The Eye of the Common Octopus," from *Frontiers in Physiology*; "Weird Pupils Let Octopuses See Their Colorful Gardens," from *UC Berkeley News*; "Do Octopuses' Arms Have a Mind of Their Own?" from *Science Daily*; "Octopuses Rewrite Their DNA to Beat the Cold," from *Science*; "Octopuses Survive

Sub-Zero Temps Thanks to Specialized Blood," from *Scientific American*; and "Touching Allows Octopuses to Pre-taste Their Food," from *ScienceNewsExplores.* There are countless other resources online that offer a compendium of fun octopus facts, such as the website Octonation.

For those interested in the cultural artifacts mentioned in this essay, the Heraklion Archaeological Museum has a terrific website that lets users view much of its collection without having to travel to Crete.

PINK HEALING

Most of the facts about fireweed included in this essay can be found in any resource about western American wildflowers. I found information specific to the plant's interactions with fire in the US Forest Service's Fire Effects Information System (FEIS) database.

PUBLICATION CREDITS

— "Tangled" was originally published in *Crazyhorse*, Fall 2020.

— "Beneath the Surface" was originally published in *Flyway*, July 11, 2024.

— "Solitude in Density" was originally published in *Tahoma Literary Review*, Issue 21.

— "When the Chestnut Falls Far" was originally published in *Chautauqua*, Summer 2024.

— "Owlgazing" was originally published in *Catamaran*, Spring 2019.

— "Negotiating Fluidity" was originally published in *The Common*, December 7, 2021.

— "Rippling Lines" was originally published in *The Florida Review*, Spring 2023.